PREFACE

In developing a large-scale simulation program for studying the effects of random errors on estimating recursively the trajectory of a spacecraft, it was necessary to generate efficiently a large sequence of random numbers. This monograph is essentially the research carried out by us in the course of generating the random numbers efficiently. These numbers were such that we were assured that their given statistical properties were known, and that any weakness in the techniques for generating these numbers had been recognized and eliminated.

We wish to acknowledge the contributions of Rosser J. Smith III, whose literature searches led to the final version of our literature review. Others who contributed by sharing their results were Dr David Falconer, Florida Technological University, A.H. Feiveson of the NASA Manned Spacecraft Center, and David Wallace of Tracor Inc.

T.G.N.

P.L.O.

McComb

PUBLISHERS' NOTE

The series of monographs in which this title appears was introduced by the publishers in 1957, under the General Editorship of Dr Maurice G. Kendall. Since that date, more than twenty volumes have been issued, and in 1966 the Editorship passed to Alan Stuart, D.Sc.(Econ.), Professor of Statistics, University of London.

The Series fills the need for a form of publication at moderate cost which will make accessible to a group of readers specialized studies in statistics or courses on particular statistical topics. Often, a monograph on some newly developed field would be very useful, but the subject has not reached the stage where a comprehensive treatment is possible. Considerable attention has been given to the problem of producing these books speedily and economically.

It is intended that in future the Series will include works on applications of statistics in special fields of interest, as well as theoretical studies. The publishers will be interested in approaches from any authors who have work of importance suitable for the Series.

CHARLES GRIFFIN & CO. LTD.

GRIFFIN'S STATISTICAL MONOGRAPHS AND COURSES

No. 1: *The analysis of multiple time-series* — M. H. QUENOUILLE
No. 2: *A course in multivariate analysis* — M. G. KENDALL
No. 3: *The fundamentals of statistical reasoning* — M. H. QUENOUILLE
No. 4: *Basic ideas of scientific sampling* — A. STUART
No. 5: *Characteristic functions** — E. LUKACS
No. 6: *An introduction to infinitely many variates* — E. A. ROBINSON
No. 7: *Mathematical methods in the theory of queueing* — A. Y. KHINTCHINE
No. 8: *A course in the geometry of* n *dimensions* — M. G. KENDALL
No. 9: *Random wavelets and cybernetic systems* — E. A. ROBINSON
No. 10: *Geometrical probability* — M. G. KENDALL and P. A. P. MORAN
No. 11: *An introduction to symbolic programming* — P. WEGNER
No. 12: *The method of paired comparisons* — H. A. DAVID
No. 13: *Statistical assessment of the life characteristic: a bibliographic guide* — W. R. BUCKLAND
No. 14: *Applications of characteristic functions* — E. LUKACS and R. G. LAHA
No. 15: *Elements of linear programming with economic applications* — R. C. GEARY and M. D. MCCARTHY
No. 16: *Inequalities on distribution functions* — H. J. GODWIN
No. 17: *Green's function methods in probability theory* — J. KEILSON
No. 18: *The analysis of variance* — A. HUITSON
No. 19: *The linear hypothesis: a general theory* — G. A. F. SEBER
No. 20: *Econometric techniques and problems* — C. E. V. LESER
No. 21: *Stochastically dependent equations: an introductory text for econometricians* — P. R. FISK
No. 22: *Patterns and configurations in finite spaces* — S. VAJDA
No. 23: *The mathematics of experimental design: incomplete block designs and Latin squares* — S. VAJDA
No. 24: *Cumulative sum tests: theory and practice* — C. S. VAN DOBBEN DE BRUYN
No. 25: *Statistical models and their experimental application* — P. OTTESTAD
No. 26: *Statistical tolerance regions: classical and Bayesian* — I. GUTTMAN
No. 27: *Families of bivariate distributions* — K. V. MARDIA
No. 28: *Generalized inverse matrices with applications to statistics* — R. M. PRINGLE and A. A. RAYNER
No. 29: *The generation of random variates* — T. G. NEWMAN and P. L. ODELL

*Now published independently of the Series.

For a list of other statistical and mathematical books see back cover.

THE GENERATION OF RANDOM VARIATES

THOMAS G. NEWMAN
Associate Professor of Mathematics
Texas Tech University

and

PATRICK L. ODELL
Professor of Mathematics and Statistics
Texas Tech University

BEING NUMBER TWENTY-NINE OF
GRIFFIN'S STATISTICAL
MONOGRAPHS & COURSES
EDITED BY
ALAN STUART, D.Sc.(Econ.)

GRIFFIN 1820 LONDON

CHARLES GRIFFIN & COMPANY LIMITED
42 DRURY LANE, LONDON, WC2B 5RX

First published 1971

ISBN: 0 85264 194 X

Set by E W C Wilkins & Associates Ltd London N12 0EH
Printed in Great Britain by Latimer Trend & Co Ltd Whitstable

CONTENTS

Chapter 1

INTRODUCTION

In this chapter we discuss briefly the history of the use of sampling techniques as an aid to solving various classes of mathematical and statistical problems. There are certain mathematical problems whose solutions are extremely difficult to obtain by the usual numerical methods, but which are relatively easily solved by sampling techniques. Sampling techniques to estimate distributions of complicated functions of random variables, such as test statistics and estimators, are important in applied statistical methodology. With the advent of operations research and systems analysis, sampling techniques have become important in performing parametric studies on conjectured models of a complicated physical system to determine performance indices over a wide range of environments.

During the early years of this century, sampling experiments were being performed to estimate the frequency distribution of various statistics, but, apart from the well-known "needle problem" [44] used to estimate the value of π, apparently very little was being done toward solving non-probabilistic problems by means of sampling techniques. However, in the 1940's sampling techniques were applied to non-probabilistic problems to gain answers to very difficult problems primarily arising in nuclear physics. The mathematicians John von Neumann and Stanley Ulam were influential in developing and applying these techniques, which became known as the Monte Carlo method. The term has survived, and common usage now includes sampling techniques used to study probabilistic and systems simulation problems, as well as deterministic problems.

The interested reader is directed to three references where the history of pertinent topics relating to the Monte Carlo method is briefly summarized, namely [100, 95, 75]. Three topics which appear fundamental to numerical simulation using sampling techniques are:

(1) modelling the problem in probabilistic form;

(2) methods to reduce the variance of the estimate of the solution; and

(3) the generation of the random variables used in the simulation.

The first topic is highly problem-related, and therefore little in general will be said in our monograph concerning this. Our aim is to investigate thoroughly the third topic, and to describe briefly the basics of the second topic insofar as they affect the third. The problem of how to generate random numbers efficiently and accurately is central to all applications of numerical simulation or of Monte Carlo techniques.

1 Early use of sampling techniques

Observations of games of chance and fitting probability density functions came much earlier than 1900 and, in fact, gave the initial stimulus to the development of probability theory. In 1906 two papers appeared, one by Weldon [108] and the second by Durbishire [24], in which dice were used for sampling purposes.

"Student" in two papers in 1908 empirically found the distribution of the standard deviation and the ratio of the mean to the standard deviation of a sample. For his random numbers he used a series of 3,000 pairs of measurements that appeared in the journal *Biometrika*. These papers are apparently the first use of sampling for estimating distribution functions.

Greenwood and White [41] used "20,000 counts" in their study of estimating a distribution from actual data. Brownlee [8] and Yule [114] used physical devices to obtain their random numbers; Brownlee tossed coins, while Yule used a circular tray with partitions into which beans were tossed while it was spinning. These simulations were performed in the 1920's.

Bispham, as reported in references [3] and [4] in 1920 and 1923, respectively, was apparently the first to sample from an arbitrary theoretical population. Bispham used a population of thirty counters, each of which had one of the integers -15 to $+15$ on it (zero excluded). A sample was obtained in order, as all thirty counters were drawn from an urn one at a time, without replacement.

Craig [16], in 1928, investigated the distribution of sample moments using punched cards. Randomness was accomplished by shuffling the cards by hand. Sheppard [93], in 1928, generated random numbers using a desk calculator. Other experiments are recorded as using physical means.

It should be noted that random numbers are still being generated using mechanical and electronic machines. The interesting electronic random number generator called "ERNIE" (Electronic Random Number Generator Indicator Equipment) was described and its properties discussed in a paper by Thomson [101] in 1959. A mechanical system using precisely manufactured rubber balls which roll about at a specified rate of speed was reported by West [109] as a means to a lottery held in the 1950's in Southern Rhodesia. The statistical properties of this apparatus are described in West's paper.

It is important to note that with the arrival of the high-speed digital computer these techniques for generating random numbers became outdated. Von Neumann [105] is said to have divided the techniques for generating random numbers into two classes: (a) physical, and (b) arithmetical. Our monograph is primarily concerned with the arithmetical technique. The physical technique involves installing a physical device as part of a computer to produce the random numbers, and this technique is not considered desirable. Interesting references on this latter topic include Brown [7] and Hamaker [46–49].

For a detailed summary of the history of distribution sampling prior to the era of the computer and its relevance to simulation and for some early references, the interested reader is referred again to Teichroew's paper [100]. For more history and a good general bibliography see the book by Tocher [103].

2 The development of Monte Carlo techniques

Dr A.S. Householder has described the Monte Carlo technique as follows: "The Monte Carlo method may briefly be described as the device of studying an artificial stochastic model of a physical or mathematical process." He goes on to state that the apparent "novelty" in the technique "lies in the suggestion that where an equation arising in a non-probabilistic context demands a numerical solution not easily obtainable by standard numerical methods, there may exist a stochastic process with distributions or parameters which satisfy the equation, and it may actually be more efficient to construct such a process and compute the statistics than to attempt to use those standard methods."

It is important to note that one must construct a stochastic process, that is to say, one must generate a sequence of random numbers whose stochastic properties are the desirable ones. It is the methods that yield these random numbers that interest us in this monograph.

In applying the Monte Carlo method to his problem, the nuclear physicist has established the method as a modern tool in nuclear physics [96]. Many references to work before 1954 are listed at the end of Meyer [75]. These references include several industrial and United States government publications, many of which are not included in our selected list of references (but see [18]).

3 Tests of random numbers

Basic to almost every use of random numbers is the capability to generate variates distributed independently uniformly on the interval [0, 1]. Once these numbers are available, one can use specified mathematical expressions which are functions of the uniform variates to obtain variates from other distributions; these latter variates can even possess specified correlation properties.

A question that arises naturally in generating uniform random numbers by any technique is: how does one become convinced that the numbers generated possess the properties desired and necessary for them to be useful in the context of a pertinent application? One usually desires that the uniform numbers in a set generated by using some technique should be –

(1) equidistributed, that is, evenly distributed over the interval [0, 1] (a precise definition of "equidistributed" can be found in Chapter 2);

(2) uncorrelated. One can minimize correlation by judiciously selecting various parameters in the mathematical expression used for generating the variates so as to minimize the absolute value of the correlation.

When someone asks, "Are the numbers generated truly random?", he is referring to (2) above. The question in (1) is essentially a check on whether or not one has reproduced numerically what one desires to have been generated.

Many people concern themselves with the obvious inconsistency that the user may claim certain sets of numbers to be random when in truth they were generated using deterministic techniques. Furthermore, the numbers can be exactly duplicated by repeating the procedure. These people prefer terms like "pseudo-random numbers" to denote these sets, especially those generated by digital computers. Certainly there is little and in most instances nothing random in the process of generating random numbers using high-speed digital computing equipment. The most that can be said about these numbers is that statistically the numbers generated can pass several valid statistical tests such as the frequency test, the serial test, the gap test, and the "poker" test [62].

Several of the more popular and perhaps more useful tests are discussed in Chapter 9 of this monograph.

4 Concluding remarks

The need for easy access to sets of random numbers became evident by 1920. Tippett [102] published the first set of tables in the *Tracts for Computers* series. Kendall and Smith [64], in 1939, published a table of 100,000 random digits. Tippett drew cards, while Kendall and Smith observed numbers from a spinning disc. Peatman and Shafer [89], in 1942, published 1,600 digits obtained from Selective Service Lists. There were others; the longest was the 1,000,000 random digits produced by a physical device at the Rand Corporation [91].

The tedious business of using tables is slow. Hence, techniques for generating pseudo-random numbers especially adaptive to high-speed digital computers have been devised and tested. These techniques are the central theme of this monograph.

There is general agreement among statisticians that sampling techniques to establish a distribution are better than nothing but are less desirable than analytical techniques. The same conclusion is, in general, true when applying sampling techniques to numerical analysis, that is, Monte Carlo techniques are inferior to direct numerical techniques. However, one can obtain reasonable solutions to extremely complicated problems using a sampling technique. It is this reason, combined with the existence of a great number of complicated problems, that has established these techniques as truly important tools in modern scientific investigations.

Chapter 2

THE GENERATION OF UNIFORM VARIATES

Many techniques for the generation of pseudo-random numbers have been suggested, tested, and used in recent years. Many of the earlier techniques were based on seemingly random phenomena. Some of these methods required rather specialized and delicate apparatus, but although they had much intuitive appeal, they often failed to produce useful numbers.

The most common techniques used in producing numerical data from random phenomena involved measurements of certain aspects of radioactive emissions or electronically generated "white noise" [59, 91]. The main complaints about such methods are based on the fact that they are slow and the nature of the apparatus does not allow a sequence to be repeated. Moreover, the lack of consistency and reliability inevitably leads to mathematical and statistical difficulties.

To alleviate many of the above difficulties, John von Neumann proposed the "mid-square method" [105]. This method is simple, iterative, and involves no special apparatus. A given sequence can be repeated whenever desired. The method requires that the user specify the number of digits in the numbers to be generated, say four, and that a starting value be specified, say 5069. The selected number is squared, yielding 25694761 in this case. The middle four digits, 6947, are selected as the first number in the sequence and are also used to generate the second number. One can thus iterate until a sufficiently large quantity of four-digit numbers have been produced. It has been noted by Lehmer [66] and others that this scheme has certain deficiencies. In the above example, numbers of the form OOAB or ABOO tend to appear and are self-perpetuating and can even eventually lead to the appearance of 0000 which terminates the sequence.

Together with the fact that a precise analysis of the statistical properties is difficult to obtain, the deficiencies mentioned above have led to the abandonment of the mid-square method in favour of other techniques which retain the favourable advantages. These methods will be presented in the next section.

1 The congruential generators

The most commonly used algorithm for the generation of pseudo-random sequences is based on calculating residues modulo some integer P of a linear transformation. Specifically, the algorithm may be written

$$R_{n+1} \equiv \lambda R_n + \mu \pmod{P}, \qquad (2.1)$$

where $0 \leqslant R_n < P$ and R_0 is a given starting value. The parameters λ and μ are integer constants, and the quality of the numbers produced depends for the most part on a suitable choice for these parameters. The modulus P is usually taken to be a function of the computer on which the computations are to be performed. On a binary computer $P = 2^b$, while on a decimal computer $P = 10^b$.

When $\mu = 0$ in (2.1), one speaks of a "pure multiplicative congruential generator" or simply a *multiplicative generator*. For $\mu \neq 0$ the procedure is called a *mixed congruential generator*.

It is customary to take the normalized sequence $R_n^* = R_n/P$ as uniformly distributed on [0,1]. The validity of this assumption can only be justified by a testing procedure on the sequence which emphasizes the aspects of randomness pertinent to a particular investigation. We defer the investigation of these tests until later (see Chapter 9) and content ourselves at present with tests relevant to the choice of parameters.

2 Period of the congruential generators

The period of a congruential generator is the maximum number of numbers which can be produced without repetition. The mixed generator can achieve a period equal to P, provided the following conditions are satisfied:

(i) g.c.d. $(\mu, P) = 1$ (g.c.d. denotes the greatest common divisor);

(ii) $\lambda \equiv 1 \pmod{q}$ for every prime factor q of P; and

(iii) $\lambda \equiv 1 \pmod{4}$ if P is a multiple of 4.

In the special case $P = 2^b$ we see that the above conditions are tantamount to $\lambda = 4\alpha + 1$ and μ odd. In the case where $P = 10^b$ we need only require that $\lambda = 20\alpha + 1$ and μ be odd and not a multiple of 5.

The above results may be found in [56], along with corresponding statements about pure multiplicative generators.

The most popular generator is the pure multiplicative generator with $P = 2^b$, since this is the most convenient choice of the modulus for a binary computer. Of course, the choice of b is dictated by the word-length of the particular computer. The maximum period in this case is 2^{b-2}, and it is not too difficult to show that

$$\text{per}(\lambda, 2^b) = 2^{b-2} \tag{2.2}$$

if and only if

$$\lambda = 8n \pm 3,$$

where we use the notation per (λ, P) to denote the period of the multiplicative generator with multiplier λ and modulus P.

Since the maximum period in (2.2) is more than adequate for most purposes, at times the choice of λ may be dictated by other considerations. In this case, it is useful to be able to determine easily per $(\lambda, 2^b)$ for a given λ. For λ even, we see that $\lambda^b \equiv 0 \pmod{2^b}$, so that $\lambda^n \equiv 0 \pmod{2^b}$ for all $n \geqslant b$. Hence per $(\lambda, 2^b)$ is at most b. Since b is usually taken to be something like $b = 35$ or $b = 48$, this case can be neglected; that is, λ should be chosen odd. From elementary number theory it is clear that per $(\lambda, 2^b)$ divides $\phi(2^b) = 2^{b-1}$, where ϕ is the Euler phi-function, that is, $\phi(n)$ is the number of positive integers less than and relatively prime to n. It follows that for λ odd there is an integer $r \geqslant 0$ such that

$$\text{per}(\lambda, 2^b) = 2^r. \tag{2.3}$$

With this result, it is a minor task to compute per $(\lambda, 2^b)$; simply calculate the residues modulo 2^b of $\lambda^2, \lambda^4, \lambda^8, \lambda^{16}, \ldots, \lambda^{2^t}, \ldots$, until the first value of t occurs with

$$\lambda^{2^t} \equiv 1 \pmod{2^b}.$$

This value of t is the r in (2.3). Note that since each term in the above sequence is obtained by squaring the preceding term, the evaluation of per $(\lambda, 2^b) = 2^r$ will require exactly r squaring operations; very conservative since $r \leqslant b-2$.

3 Linear transformations of uniform variates

Let the notation $U \sim U(a, b)$ mean that the random variable U is uniformly distributed on the interval $[a, b]$. If the linear transformation

$$T(x) = (b-a)x + a, \quad a < b \tag{2.4}$$

is applied to a random variable $U \sim U(0, 1)$, it follows that $T(U) \sim U(a, b)$. Moreover, if $b-a$ is an integer, then the "fractional part," denoted by $\{T(U)\}$, of $T(U)$ is again uniformly distributed on $[0, 1]$. If we let $[x]$ be the greatest integer not exceeding x, then the transformation

$$S(x) = \{T(x)\} \equiv T(x) \pmod 1$$

may be written as

$$S(x) = T(x) - [T(x)]. \tag{2.5}$$

It is convenient to let $\lambda = b-a$ and $\mu = a$ in (2.4), so that it becomes

$$T(x) = \lambda x + \mu, \quad 0 \leqslant x \leqslant 1. \tag{2.6}$$

Finally, we note that since $S(x)$ is the residue of $T(x)$ modulo 1, we have $0 \leqslant S(x) < 1$, where

$$S(x) \equiv \lambda x + \mu \pmod 1. \tag{2.7}$$

Now the form of (2.7) is identical with that of (2.1). In fact, with λ and μ taken from (2.1), the sequence

$$R^*_{n+1} = \{\lambda R^*_n + \mu/P\} \tag{2.8}$$

obtained from (2.7) is identical with the normalized sequence R_n/P produced by (2.1), provided only that $R^*_0 = R_0/P$. In other words, the properties of the normalized sequences produced by a congruential generator may be studied by investigating transformations defined by (2.7), or equivalently (2.5) and (2.6).

We now proceed to derive the correlation coefficient between U and $S(U)$. The results can then be applied to the congruential generators. Since we want $S(U) \sim U(0,1)$, we take λ to be an integer, and with no loss in generality we assume $\lambda > 0$. Also, from (2.7) we see that it is sufficient to assume $0 \leqslant \mu < 1$; otherwise, we may replace μ by $\{\mu\}$. We state the results in the following

Theorem 2.1 If λ is a positive integer and $0 \leqslant \mu < 1$, then the correlation coefficient ρ between U and $T(U)$, when $U \sim U(0,1)$, is given by

$$\rho(U, S(U)) = \frac{1 - 6\mu(1-\mu)}{\lambda}. \tag{2.9}$$

Proof The major task involved is the evaluation of the integral

$$\begin{aligned} E(US(U)) &= \int_0^1 xS(x)\,dx \\ &= \int_0^1 x(\lambda x + \mu - [\lambda x + \mu])\,dx \\ &= \lambda \int_0^1 x^2\,dx + \mu \int_0^1 x\,dx - \int_0^1 x[\lambda x + \mu]\,dx \\ &= \lambda/3 + \mu/2 - \int_0^1 x[\lambda x + \mu]\,dx. \end{aligned}$$

Integrating the last integral by parts, we obtain

$$E(US(U)) = \mu/2 - \lambda/6 + \frac{1}{2}\int_0^1 x^2\,d[\lambda x + \mu], \tag{2.10}$$

where the last integral is a Stieltjes integral. The discontinuities of $[\lambda x + \mu]$, for $0 \leqslant x \leqslant 1$, occur when $\lambda x + \mu = k$, an integer, $1 \leqslant k \leqslant \lambda$. Therefore,

$$\frac{1}{2}\int_0^1 x^2 d[\lambda x + \mu] = \frac{1}{2}\sum_{k=1}^{\lambda}\left(\frac{k-\mu}{\lambda}\right)^2$$

$$= \frac{1}{6}\lambda + \frac{1}{4} + \frac{1}{12\lambda} - \frac{\mu}{2} - \frac{\mu}{2\lambda} + \frac{\mu^2}{2\lambda}. \qquad (2.11)$$

Now from (2.10) and (2.11) we have

$$E(US(U)) = \frac{1}{4} + \frac{1}{12\lambda} - \frac{\mu}{2\lambda} + \frac{\mu^2}{2\lambda}$$

$$= \frac{1}{4} + (1 - 6\mu(1-\mu))/12\lambda.$$

We may now evaluate the covariance of U and $T(U)$ and obtain

$$\text{cov}(U, S(U)) = \frac{1 - 6\mu(1-\mu)}{12\lambda}.$$

Finally, since U and $S(U)$ each have variance equal to $\frac{1}{12}$, it follows that the correlation coefficient is given by

$$\rho(U, S(U)) = \frac{1 - 6\mu(1-\mu)}{\lambda},$$

which was to be shown.

An interesting result, but not a very useful one, may be obtained by equating $1 - 6\mu(1-\mu)$ to zero. The roots are

$$\mu = \frac{1}{2} \pm \frac{\sqrt{3}}{6},$$

and for these choices of μ the correlation is zero.

4 Correlation of congruential generators

With the aid of Theorem 2.1 and equation (2.8) we can obtain an estimate for the correlation between successive terms in the sequence generated by a congruential generator of the type given in (2.1). Replacing μ by μ/P in equation (2.9), we obtain

$$\rho = \frac{1 - 6(\mu/P)(1 - \mu/P)}{\lambda}. \qquad (2.12)$$

This result agrees with that given by R.R. Coveyou [15]. Unfortunatel its validity is weak since it involves some rather strict assumptions concerning discrete distributions. The error is not too great, however. Greenberger [38] has given a derivation which shows that (2.12) needs a correction factor ϵ, where $|\epsilon| < \lambda/P$. At least we can see that values of λ which are very small or very large (relative to P) can give trouble.

In most instances it is desirable to control the correlation between several consecutive terms. A simple exercise in induction gives the following result:

Theorem 2.2 If $R_{n+1} \equiv \lambda R_n + \mu \pmod P$, as in (2.1), then, for $k \geqslant 1$,

$$R_{n+k} \equiv \lambda_k R_n + \mu_k \pmod P, \qquad (2.13)$$

where

$$\lambda_k \equiv \lambda^k \pmod P, \quad 1 \leqslant \lambda_k < P$$

and

$$\mu_k \equiv \frac{\lambda^k - 1}{\lambda - 1}\,\mu \pmod P, \quad 0 \leqslant \mu_k < P.$$

Evidently, in view of (2.12) and the remarks which followed, it is desirable to choose λ so that the successive residues $\lambda_1, \lambda_2, \ldots, \lambda_k$ of powers of λ become neither too large nor too small, relative to P. One way to accomplish this is to choose λ so that λ_2 is approximately equal to λ. For various reasons, including the above as well as machine convenience, in the case where $P = 2^b$ the choice $\lambda = 2^a + 3$, $a = [b/2] + 1$, has been recommended by Moshman [79]. For $b = 35$ this gives $\lambda = 2^{18} + 3$. For this choice, the sequence $\lambda_2, \lambda_3, \ldots$, is well behaved for the first few terms. Note that this results in a value of λ of approximately $\sqrt{P}$. Thus, even with Greenberger's correction term, the correlation coefficient is of the order of $1/\sqrt{P}$.

5 Equidistribution and white sequences

A discussion of pseudo-random numbers would not be complete without outlining some of the known results of a more theoretical nature concerning properties of deterministic sequences. We will concern ourselves with an infinite sequence from the interval [0, 1] and study the properties of such sequences which are characteristic of random sequences.

The most obvious requirement of a random sequence is that it should be equidistributed in [0, 1]. Precisely:

Definition A sequence $\{x_n\}$ is equidistributed in [0, 1] if for each a and b with $0 \leqslant a < b \leqslant 1$ we have

$$\lim_{n\to\infty} \frac{K_n}{n} = b - a,$$

where K_n is the number of times that $a \leqslant x_i < b$ for $i \leqslant n$.

More generally, if $\{z_n\}$ is a sequence of k-dimensional vectors in the k-dimensional unit cube, that is

$$z_n = (x_n^{(1)}, x_n^{(2)}, \ldots, x_n^{(k)}), \quad 0 \leqslant x_n^{(i)} \leqslant 1,$$

then we say that $\{z_n\}$ is equidistributed, provided that whenever $0 \leqslant a_i < b_i \leqslant 1$, then

$$\lim_{n\to\infty} \frac{K_n}{n} = (b_1 - a_1) . (b_2 - a_2) \ldots (b_k - a_k),$$

where K_n is the number of times that z_i falls in the rectangle

$a_1 \leqslant x^{(1)} \leqslant b_1, \ldots, a_k \leqslant x^{(k)} \leqslant b_k$ for $i \leqslant n$.

By specifying a starting value x_0 one can define a sequence by

$$x_{n+1} = S(x_n), \tag{2.14}$$

where S is as defined by (2.7). In [29] the resulting sequence is shown to be equidistributed for almost all x_0. Apparently this is not as strong a result as it might appear, since there is not a single *known* value of x_0 which yields an equidistributed sequence.

From any sequence $\{x_n\}$ we can form a k-dimensional sequence by taking successive groups of k terms and defining the k-dimensional derived sequence by

$$Z_n = (x_{nk}, x_{nk+1}, \ldots, x_{nk+k-1}).$$

For certain sequences this can lead to a curious effect, as evidenced by the following result:

Theorem 2.3 If the sequence $\{x_n\}$ is given by a recursion formula

$$x_{n+1} = F(x_n),$$

and if F has a point of continuity in $(0, 1)$, then for every $k > 1$, the k-dimensional derived sequence is not equidistributed.

Proof Suppose F is continuous at the point a. Choose $\delta > 0$ so that for $|x - a| < \delta$ we have $0 \leqslant x \leqslant 1$ and $|F(x) - F(a)| < \frac{1}{4}$. Let F be any sub-interval of $[0, 1]$ whose distance from $F(a)$ exceeds $\frac{1}{4}$ and let I be the set of x with $|x - a| < \delta$. Then no point $(x, F(x))$ lies in the rectangle $I \times J$. It follows that no two-dimensional derived sequence, and consequently no k-dimensional derived sequence, is equidistributed regardless of the initial value x_0.

The above result, as well as others to be cited, is due to Franklin [30]. Note that the theorem applies to the sequence defined by (2.14), since S has at most a finite number of points at which discontinuities occur. Likewise, the result applies to the congruential generators. In fact, Greenberger [40] has reported on just such an effect. Namely, in the sequence R_n^* produced by a congruential generator there are narrow bands in the (R_n^*, R_{n+1}^*) plane which are completely devoid of points.

For an equidistributed sequence $\{X_n\}$ we define the autocorrelation function, provided the sum exists, by

$$R(k) = \sum_{n=1}^{\infty} \left(x_n - \tfrac{1}{2}\right)\left(x_{n+k} - \tfrac{1}{2}\right). \tag{2.15}$$

The sequence $\{x_n\}$ is said to be a "white" sequence, provided $R(k) = 0$ for $k = 1, 2, \ldots$. "Whiteness" is thus in a sense a measure of independence among the terms in the sequence.

As in Theorem 1.2 we see that if a sequence $\{x_n\}$ is generated from (2.14), then we can write

$$x_{n+k} = \{\lambda^k x_n + \mu_k\}, \tag{2.16}$$

where $\mu_k = \{\mu(\lambda^k - 1)/(\lambda - 1)\}$ (here $\{.\}$ again denotes "the fractional part of"). From equation (2.9) we see that for this sequence,

$$R(k) = \frac{1 - 6\mu_k(1 - \mu_k)}{12\lambda^k}.$$

Hence $R(k) \to 0$ as $k \to \infty$. In this sense this sequence is asymptotically white. Note that this is contingent upon x_0 being chosen so that $\{x_n\}$ is equidistributed.

It should be clear by this time that the sequences produced by congruential generators suffer from defects which for many applications could be disastrous. It is hard to give a precise rule as to when one may expect trouble from a congruential generator; indeed, they are certainly adequate for most applications. At times, however, the assumption of independence of successive random numbers may be crucial. For instance, the simulation of stationary normal processes presented in Chapter 6 depends strongly upon starting with a white sequence to generate independent normal variates. In these cases care must be taken to generate a sequence which is reasonably white.

In [30] it is shown that for almost all $\theta > 1$, the sequence

$$x_n = \{\theta^n\} \tag{2.17}$$

is a white sequence, and in fact has other properties desirable in a random sequence. In [32] and [74] the actual generation of a sequence from (2.17) with $\theta = \pi$ is discussed. Since the actual computations are dependent upon available computing machinery we shall not present any further details here.

6 Summary

For most applications the most suitable generators of uniform variates seem to be the pure multiplicative generators. The results given indicate that sequences which are sufficiently long can easily be generated by these methods, and the statistical characteristics are good enough for most purposes. The recommendations for the choice of parameters are meant only to serve as guide-lines to help reject bad choices rather than find good ones. In practice one should design tests of these sequences based on the specific applications. Some more or less standard tests will be discussed in Chapter 9.

If the particular needs of a problem cannot be met by a multiplicative generator it is recommended that one should compute a sequence of the form

$$x_n = \{\theta^n\}, \quad \theta > 1$$

as accurately as possible. Again, specific tests for the resulting sequences must be established and a search for a suitable sequence carried out.

Chapter 3

GENERATING NORMAL VARIATES

1 The central limit theorem

The normal distribution plays a fundamental role in the theory of statistics. A random variable X having this distribution, with mean μ and variance σ^2, has the density function

$$f(x) = \frac{1}{\sigma\sqrt{2\pi}} \exp\left[-(x-\mu)^2/(2\sigma^2)\right], \quad -\infty < x < \infty. \tag{3.1}$$

The standardized normal distribution is the special case $\mu = 0$, $\sigma^2 = 1$. In the case where X is standardized normal, the random variable $Z = \sigma X + \mu$ is normal with mean μ and variance σ^2. Thus, for purposes of random number generation there is no loss in generality when one specializes to the standardized normal.

Perhaps the most expedient means of producing normal variates is by appealing to the central limit theorem [76, p.149]. In the case which is of interest we may interpret this theorem as follows:

Suppose $X_1, X_2, \ldots$ are independent random variates with the same density function $f(x)$ and with finite mean μ and variance σ^2. For each positive integer n, let

$$Z_i = \frac{1}{n} \sum_{i=1}^{n} X_i \tag{3.2}$$

and let

$$Y_n = \frac{Z_n - \mu}{\sigma} \sqrt{n} \,. \tag{3.3}$$

Then the density of Y_n approaches the standardized normal as $n \to \infty$. Thus for large sample sizes one can produce standardized normal variates from any variates X with mean μ and variance σ^2. In the case where X is uniform on $[0,1]$ the transformation becomes

$$Y_n = \frac{Z_n - \frac{1}{2}}{1/\sqrt{12}} \sqrt{n} \tag{3.4}$$

$$= \sqrt{3/n}\left(2 \sum_{i=1}^{n} u_i - n\right),$$

where $u_1, u_2, \ldots, u_n$ is a sample of size n.

It has been observed [82] that the natural choice $n = 12$ is too small to yield a reasonable approximation. On the other hand, for $n \geqslant 50$ the central limit theorem is relatively insensitive to the form of $f(x)$. Moreover, the time required to produce a single normal variate is comparatively long – of the order of 5 milliseconds for the IBM 704.

An additional disadvantage of the method, particularly when using a uniform X, is that the tails of the normal distribution will not be well represented. Thus, if a particular investigation depends on sampling the tails, then the above method is unsatisfactory. However, for those who require only a small sample of normal variates and for whom the above objections do not apply, the method can prove quite adequate.

2 The Box–Muller method

Provided one's computing machinery can accurately evaluate square roots, logarithms, and trigonometric functions in a reasonable length of time, there is a very attractive method for generating normal variates which was suggested by Box and Muller [6]. It is summarized in the following result:

Theorem 2.1 If U and V are independent random variables, both distributed uniformly on [0,1], then

$$X = (-2 \ln U)^{1/2} \cos 2\pi V \tag{3.5a}$$

and

$$Y = (-2 \ln U)^{1/2} \sin 2\pi V \tag{3.5b}$$

are independent random variables with the standardized normal distribution.

3 Decomposition method

The method to be discussed here is quite general and applies equally well to any distribution. However, following the general treatment, we shall illustrate the procedure for the normal distribution.

Suppose we are given random variables X_1, X_2, ..., and a sequence a_1, a_2, ... of positive constants satisfying

$$\sum_{n=1} a_n = 1.$$

Let us define a new random variable X by equating X to X_n with probability a_n. Thus X is a mixture of X_1, X_2, ... Now if X_n has a distribution function $F_n(x)$ then the distribution function for X is easily seen to be

$$F(x) = \sum_n a_n F_n(x), \tag{3.6}$$

since $F(x) = \Pr(X \leqslant x) = \sum_n a_n \Pr(X_n \leqslant x)$.

From (3.6) it also follows that if X_n has the probability density function $f_n(x)$ then X has the probability density function

$$f(x) = \sum_n a_n f_n(x) \tag{3.7}$$

Now if we have available a collection of random number generators α_n, one for each X_n, then we can generate values of X by using α_n with probability a_n. Of course, this involves an additional generator α_0 to generate the discrete distribution of the subscript n.

Before proceeding further let us remark that if X_1, X_2, ... all have the same distribution $G(x)$, then for any choice of the a_n we have $F(x) = G(x)$.

It has been observed several times (see [68]) that the uniform number generators discussed in Chapter 2 can be improved by a mixing process. It is customary to use m generators and assign $a_1 = a_2 = \ldots = a_m = 1/m$. However, other choices of the a_n can be made.

The idea now is to express a random variable X which is perhaps difficult to generate as a mixture of other random variables which are easily generated. It is convenient to work with equation (3.7). Suppose we are given $f(x)$. If we have a probability density function $g(x)$ and a positive number a such that, for all x,

$$f(x) - ag(x) \geqslant 0, \tag{3.8}$$

then it follows that

$$1 - a = \int_{-\infty}^{\infty} (f(x) - ag(x)dx > 0, \tag{3.9}$$

unless perhaps $g(x) = f(x)$. Thus, with

$$0 < 1 - a \quad \text{and} \quad a > 0$$

we have

$$0 < a < 1.$$

Now let

$$h(x) = \frac{f(x) - ag(x)}{1 - a} \tag{3.10}$$

and note that $h(x)$ is a probability density function. Moreover,

$$f(x) = ag(x) + (1 - a)h(x), \tag{3.11}$$

so that we have a decomposition of X as a mixture.

Now if the distribution corresponding to $g(x)$ is easily generated, it is desirable that a be as large as possible. Thus, we choose

$$a = \max \{\lambda \mid f(x) - \lambda g(x) \geqslant 0 \quad \text{for all } x\}. \tag{3.12}$$

Since $f(x) - g(x)$ is continuous with respect to λ, a clearly exists and satisfies $f(x) - ag(x) \geqslant 0$ for all x. From equation (3.9) we have $a < 1$. The problem is now reduced to decomposing $h(x)$, and to do this we may apply the same procedure.

We now describe the general procedure. Given the density function $f(x)$ of X, we choose a density function $f_1(x)$ satisfying the conditions of equation (3.8). We set

$$a_1 = \max \{\lambda \mid f(x) - \lambda f_1(x) \geqslant 0 \quad \text{for all } x\}.$$

Having constructed $f_1, f_2, \ldots, f_{n-1}$ we choose a density function f_n for which a positive number λ exists, with

$$f(x) - a_1 f_1(x) - \ldots - a_{n-1} f_{n-1}(x) - \lambda f_n(x) \geqslant 0 \tag{3.13}$$

for all x. We then let a_n be the maximum λ satisfying (3.13). Now if

$$f(x) - \sum_{i=1}^{n} a_i f_i(x) > 0 \quad \text{for some } x, \tag{3.14}$$

then

$$b_n = \sum_{i=1}^{n} a_i < 1.$$

We compute the density function $h_n(x)$ of the residual (i.e. the error in approximating X) by normalizing the difference in (3.14). Thus,

$$h_n(x) = \frac{f(x) - \sum_{i=1}^{n} a_i f_i(x)}{1 - \sum_{i=1}^{n} a_i} \tag{3.15}$$

We have

$$f(x) = \sum_{i=1}^{n} a_i f_i(x) + b_n h_n(x), \tag{3.16}$$

and in terms of distributions,

$$F(x) = \sum_{i=1}^{n} a_i F_i(x) + b_n H_n(x). \tag{3.17}$$

In practice one can usually choose the a_j so that b_n becomes small very rapidly. We shall now illustrate this by approximating the standardized normal distribution. Hence,

$$f(x) = \frac{1}{\sqrt{2\pi}} e^{-x^2/2}$$

Let

$$X_1 = 2(U_1 + U_2 + U_3 - 1{\cdot}5), \tag{3.18}$$

where U_1, U_2, and U_3 are independently distributed $U(0,1)$. The density function for X_1 is given by

$$f_1(x) = \begin{cases} (3 - x^2)/8, & -1 \leqslant x \leqslant 1 \\ (3 - |x|)^2/16, & 1 \leqslant |x| \leqslant 3 \\ 0 & |x| \geqslant 3 \end{cases} \tag{3.19}$$

Finding a_1 is done by finding the point at which the quotient $f(x)/f_1(x)$ is minimized. Zeros of the derivative occur at $|x| = 0,1,2$. The minimum value of the ratio occurs for $x = \pm 2$. Thus,

$$a_1 = \frac{f(2)}{f_1(2)} = \frac{16e^{-2}}{\sqrt{2\pi}} \simeq \cdot 8638554642. \tag{3.20}$$

Note that equation (3.18) can be used to generate X more than 86 per cent of the time.

It is not hard to see that the residual density function $h_1(x)$ is well approximated by

$$f_2(x) = \begin{cases} (6-4x)/9, & |x| \leqslant 1{\cdot}5 \\ 0, & |x| > 1{\cdot}5 \end{cases} \tag{3.21}$$

which is the density function of

$$X_2 = 1{\cdot}5(U_1 + U_2 - 1). \tag{3.22}$$

Resorting to a numerical method, it can be found that the ratio $(f(x) - a_1 f_1(x))/f_2(x)$ has a minimum value of

$$a_2 = \cdot 110817965$$

occurring at $|x| = \cdot 87386312884$. Now (3.22) can be used to generate X an additional 11 per cent of the time.

So far we have obtained samples from the portion of the normal curve between $x = -3$ and $x = +3$. Moreover, we still have a residual error in this interval since $f(x) - a_1 f_1(x) - a_2 f_2(x)$ is not identically zero in the interval $[-3,3]$. Before dealing with this error, let us eliminate the error due to failure to sample from the tails of $f(x)$ described by $|x| \geqslant 3$.

The probability that $|X| \geqslant 3$ is given by

$$a_3 = 2\int_{-\infty}^{-3} f(x)\,dx \simeq \cdot 002699796063, \tag{3.23}$$

which occurs less than $\cdot 3$ per cent of the time. Let $f_3(x)$ be the conditional density function of X, given $|X| \geqslant 3$. The form of $f_3(x)$ is easily obtained, but since we shall not need to evaluate f_3, we omit it.

We propose to sample X_3 by means of the Box–Muller method (see section 2, page 19) and a rejection technique. Notice that in equations (3.5), X and Y have magnitude not exceeding $(-2 \ln U)^{1/2}$. Since we are sampling the tails we must reject all values of X and Y for which $(-2 \ln U)^{1/2} \leqslant 3$. It follows that the maximum value of U for which X or Y could be accepted is

$$\lambda_0 = e^{-9/2} .$$

Now if $U \sim U(0, 1)$, then $\lambda_0 U \sim U(0, \lambda_0)$. We replace U by $\lambda_0 U$ in equations (3.5) and obtain

$$\left.\begin{aligned} X &= (9 - 2 \ln U)^{1/2} \cos 2\pi V \\ \text{and} \qquad Y &= (9 - 2 \ln U)^{1/2} \sin 2\pi V, \end{aligned}\right\} \tag{3.24}$$

where $U \sim U(0, 1)$ and $V \sim U(0, 1)$. Now to obtain a sample of X_3 we generate values of U and V and calculate successively the values of X and Y from (3.24), accepting the first value which is outside the interval $[-3, 3]$. The use of the selective method of sampling given above greatly improves the selection rate, namely by the factor of $\lambda_0^{-1} = e^{9/2}$, over that of crude rejection whereby pairs are generated directly from equations (3.5).

Now at this stage, the residual

$$h_3(x) = \frac{f(x) - a_1 f_1(x) - a_2 f_2(x) - a_3 f_3(x)}{1 - a_1 - a_2 - a_3}$$

is all that is left to make our method exact. Let us take

$$f_4(x) = h_3(x) \quad \text{and} \tag{3.25}$$

$$a_4 = 1 - a_1 - a_2 - a_3 \simeq \cdot 02262677245.$$

Observe that we have represented $f(x)$ exactly; that is, $h_4(x) = 0$.

Now $f_4(x)$ is seen to be zero for $|x| \geqslant 3$, since we handled the tails exactly by means of $f_3(x)$. It can be determined that the function $f_4(x)$ has a maximum value of approximately $\cdot 3181471173$. Therefore, if

$$U_1 = 6U - 3 \tag{3.26}$$

and

$$V_1 = \cdot 3181471173\, V,$$

then (U_1, V_1) is uniformly distributed over the rectangle enclosing the relevant portions of $f_4(x)$. Moreover, of the points (u_1, v_1) comprising this rectangle, u_1 can be taken as a variate with density $f_4(u_1)$ if and only if $v_1 \leqslant f_4(u_1)$. Hence, to generate values of X_4 we generate pairs (u_1, v_1) until we encounter a pair with $v_1 \leqslant f_4(u_1)$. We then set $X_4 = u_1$.

The area of the rectangle above is about $A = 1{\cdot}908882704$. Hence the probability that a pair (u_1, v_1) will be accepted is $A^{-1} = {\cdot}5238666566$; that is, we expect to need about two pairs of uniform variates to produce a value of X_4.

To compare this with the acceptance rate for values of X_3, let us compute the probability p that at least one of (X, Y) produced by the Box–Muller method will be accepted (crude acceptance rate). This is given by

$$p = 1 - (1 - a_2)^2 = {\cdot}00439.$$

For the refined sampling method above, the acceptance rate is

$$pe^{9/2} = {\cdot}39517,$$

since $\lambda_0^{-1} = e^{9/2} = 90{\cdot}017$.

As a final word let us summarize the steps involved.

(I) With probability $a_1 = {\cdot}8638554642$, let

$$X = 2(u_1 + u_2 + u_3 - 1{\cdot}5).$$

(II) With probability $a_2 = {\cdot}110817965$, let

$$X = 1{\cdot}5(u_1 + u_2 - 1).$$

(III) With probability $a_3 = {\cdot}002699796063$, generate pairs

$$X_1 = (9 - 2 \ln u_1)^{1/2} \cos 2\pi u_2$$

$$X_2 = (9 - 2 \ln u_1)^{1/2} \sin 2\pi u_2$$

and accept the first value outside $[-3, 3]$.

(IV) With probability $a_4 = {\cdot}02262677245$, generate pairs (x, y) until $y \leqslant f_4(x)$, and then let $X = x$, where

$$x = 6u_1 - 3$$

$$y = {\cdot}3181471173\, u_2$$

and

$$f_4(x) = \begin{cases} ae^{-x^2/2} - b(3 - x^2) - c(1{\cdot}5 - |x|), & |x| < 1 \\ ae^{-x^2/2} - d(3 - |x|)^2 - c(1{\cdot}5 - |x|), & 1 \leqslant |x| < 1{\cdot}5 \\ ae^{-x^2/2} - d(3 - |x|)^2, & 1{\cdot}5 \leqslant |x| < 3 \\ 0, & |x| \geqslant 3, \quad \text{and} \\ a = 15{\cdot}75192787 & c = 1{\cdot}944694161 \\ b = 4{\cdot}263583239 & d = 2{\cdot}1317916185. \end{cases}$$

The decomposition method given above is very appealing because of its speed. If the time needed to carry out step i is t_i then the expected amount of time to produce a single variate is given by

$$t_e = t_0 + \sum_{i=1}^{n} a_i t_i ,$$

where t_0 is the time needed to generate the discrete distribution of i. Of course, the value of t_e is a function of a particular computer, but the merit of the method is that it involves doing an easy computation often and a difficult one rarely.

For references to similar concepts the reader is referred to [53] and [57].

Chapter 4

DISTRIBUTIONS OBTAINED FROM THE UNIFORM AND THE NORMAL

1 The chi-square distribution, $\chi^2(n)$

If X is a random variable having the chi-square distribution with n degrees of freedom, then the probability density function of X is given by

$$f(x) = \frac{x^{(n/2)-1}\ e^{-x/2}}{(n/2)2^{n/2}} \tag{4.1}$$

where $x > 0$ and $n = 1, 2, \ldots$. This distribution has several elementary properties which aid in its generation. If X and Y are independently distributed as chi-square with n and m degrees of freedom, respectively, then $X + Y$ has the chi-square distribution with $n + m$ degrees of freedom. If $X_1, X_2, \ldots, X_n$ are independently $N(0, 1)$, then $Y = X_1^2 + X_2^2 + \ldots + X_n^2$ has a chi-square distribution with n degrees of freedom. These facts may be found in the standard statistics texts such as [76] or [110].

From equations (3.5) and the last remark above, it follows that if U is distributed $U(0, 1)$, then

$$X = -2 \ln U \tag{4.2}$$

has a chi-square distribution with two degrees of freedom. More generally, by the additive property of the chi-square, if $U_1, U_2, \ldots, U_m$ are independently distributed $U(0, 1)$ then

$$X = -2 \ln (U_1\ U_2 \ldots U_m) \tag{4.3}$$

has a chi-square distribution with $2m$ degrees of freedom. Moreover, if Y is independent of X and is distributed $N(0, 1)$, then

$$Z = X + Y^2 \tag{4.4}$$

has a chi-square distribution with $n = 2m + 1$ degrees of freedom.

The equations (4.3) and (4.4) allow the generation of the chi-square distribution with any desired degree of freedom.

Chi-square variates with n degrees of freedom can be produced accurately and conveniently for large n by means of the so-called Wilson–Hilferty transformation [111] and [73]). If X is distributed as chi-square with n degrees of freedom, then

$$Y = (9n/2)^{1/2}[(X/n)^{1/3} + 2/(9n) - 1] \tag{4.5}$$

has a standardized normal distribution. The error in (4.5) was studied numerically by Mathur [73]. The maximum absolute error $|E|$ for $n \geqslant k$ between the density function of Y and the standardize normal is as follows:

k	$\|E\|$
1	·03443
2	·01218
3	·00692
5	·00353
10	·00148
12	·00119
13	·00109
14	·00103
15	·00092

Solving equation (4.5) for X, we obtain the Wilson–Hilferty transformation

$$X = n[Y\sqrt{2/(9n)} + 2/(9n) + 1]^3, \tag{4.6}$$

where Y is distributed $N(0, 1)$. The study by Mathur shows that (4.6) is more reliable, especially for smaller values of n, than the more familiar transformation

$$X = \tfrac{1}{2}(Y - \sqrt{2n - 1})^2,$$

which is often used for $n > 25$.

2 The exponential distribution

The distribution function for the exponential distribution with mean $\frac{1}{2}$ is given by

$$F(x) = 1 - e^{-\lambda x}.$$

If $U \sim U(0,1)$, then by equating $F(x)$ to $1 - U$ we see that

$$x = -(1/\lambda) \ln U \tag{4.7}$$

has the above exponential distribution.

3 The logistic distribution, $\mathfrak{L}(\alpha, \beta)$

The distribution function for the logistic distribution with parameters α and β, $\beta > 0$, is given by

$$F(x) = (1 + \exp[-(x - \alpha)/\beta]).$$

Proceeding as with the exponential distribution, it follows that

$$X = \alpha - \beta \ln[(1 - U)/U] \tag{4.8}$$

has the logistic distribution $\mathfrak{L}(\alpha, \beta)$.

4 The Gamma distribution, $\Gamma(\lambda, k)$

If the random variable X has a Gamma distribution with parameters λ and k, then X has density function

$$f(x) = \frac{\lambda}{\Gamma(k)} (\lambda x)^{k-1} e^{-\lambda x}, \tag{4.9}$$

where $x \geqslant 0$, $k > 0$, and $\lambda > 0$. Note that for $k = n/2$ and $\lambda = \frac{1}{2}$, this becomes the chi-square distribution with n degrees of freedom. The above distribution can be characterized by its moment-generating function [76, p. 129], where $t < \lambda$,

$$M_X(t) = \mathrm{E}[e^{Xt}] = \left(1 - \frac{t}{\lambda}\right)^{-k}. \tag{4.10}$$

We recall that if Y is a random variable having the exponential distribution with mean $\mathrm{E}(Y) = 1/\lambda$, then Y has density function

$$h(y) = \lambda e^{-\lambda y},$$

where $y \geqslant 0$ and $\lambda > 0$. This distribution is characterized by the following moment-generating function [76, p. 119]:

$$M_Y(t) = \left(1 - \frac{t}{\lambda}\right)^{-1}.$$

If Y_1, Y_2, ..., Y_k are k independent random variables, each having density function $h(y)$ above, then the moment-generating function for

$$X = \sum_{i=1}^{k} Y_i$$

is given by [76, p. 121] :

$$M_X(t) = \left(1 - \frac{t}{\lambda}\right)^{-k},$$

which is exactly (4.10). Consequently, this X has density function (4.9).

Now if U_i has the uniform distribution over [0, 1], then

$$Y_i = -\frac{1}{\lambda} \ln U_i$$

has the exponential distribution with density function $h(y)$ above. Thus, if $U_1, U_2, \ldots, U_k$ are k independent random variables, each having the uniform distribution over [0, 1], it follows that

$$X = \sum_{i=1}^{k} Y_i = -\frac{1}{\lambda} \ln \prod_{i=1}^{k} U_i \qquad (4.11)$$

is a gamma variate with parameters λ and k.

5 The beta distribution, $\beta(p, q)$

The density function for the beta distribution [76] with parameters p and q is given by

$$f(x) = \frac{\Gamma(p+q+2)}{\Gamma(p+1)\,\Gamma(q+1)} x^p (1-x)^q,$$

where $0 \leqslant x \leqslant 1$, $p > -1$, $q > -1$.

In the case where p and q are integers, there is a well-known method for obtaining the desired beta variate from two independent gamma variates. If $X_1 \sim \Gamma(1, p)$ and $X_2 \sim \Gamma(1, q)$ then

$$Y = X_1/(X_1 + X_2)$$

has the desired beta distribution [76].

In general, it is possible to generate $\beta(p, q)$ by means of a rejection method. Let U and V be independently distributed $U(0, 1)$ and define new random variables S and T by

$$S = [\Gamma(p+2)U/\Gamma(p+q+3)]^{1/(p+1)}$$
$$T = [\Gamma(q+2)V]^{1/(q+1)}. \tag{4.12}$$

It can be shown that, subject to the restraint $S + T \leqslant 1$, the random variable

$$X = S/(S+T)$$

has the desired beta distribution. Thus, to generate values of X, we generate values s and t of S and T until

$$s + t \leqslant 1$$

and take

$$X = x = s/(s+t). \tag{4.13}$$

6 The *F* distribution

If X is a random variable having the F distribution, with m and n degrees of freedom, then X has the density function

$$f(x) = \frac{\Gamma((m+n)/2)}{\Gamma(m/2)\Gamma(n/2)}(m/n)^{m/2}\frac{x^{(m-2)/2}}{(1+mx/n)^{(m+n)/2}}$$

where $x \geqslant 0$, $m > 0$, and $n > 0$. (In practice, m and n are positive integers.)

There is a well-known transformation [76, p. 244] :

$$Y = \frac{mX/n}{1+mX/n}$$

where Y is a beta variate with parameters $p = (m/2) - 1$ and $q = (n/2) - 1$, and X has the F distribution, with m and n degrees of freedom.

Since a method for generating beta variates has been given, it is convenient to rewrite the above as

$$X = nS/mT,$$

subject to $S + T \leqslant 1$.

Thus, to generate X we generate values s and t of S and T from equations (4.12) until $s + t \leqslant 1$ and then take

$$X = x = ns/mt. \tag{4.14}$$

7 Student's *t* distribution

If X is a random variable having Student's t distribution with n degrees of freedom, then X has density function

$$f(x) = \frac{\Gamma((n+1)/2)}{\sqrt{2\pi}\,\Gamma(n/2)} (1 + x^2/n)^{-(n+1)/2}$$

where $-\infty < x < \infty$, and $n > 0$ (in practice, n is usually a positive integer) [76, p. 233].

If X has Student's t distribution with n degrees of freedom, then X^2 has the F distribution with one and n degrees of freedom [76, *ibid.*]. Therefore, if Y has this F distribution, then the desired t variate X can be generated by

$$X = \pm\sqrt{Y}, \tag{4.15}$$

where the sign is chosen at random with equal probabilities.

8 The acceptance rate for the beta distribution

We recall that the three distributions in this section are derived from the beta distribution, and each of the generation methods essentially requires that $s + t \leqslant 1$. Consequently one would desire the probability that a point (u_1, u_2) would provide a valid variate. For the case of the beta variate this probability is

$$\Pr[s + t \leqslant 1] = \Pr[u_1 \leqslant w]$$

$$= \int_0^1 \int_0^w du_1\, du_2 = \int_0^1 w\, du_2,$$

where

$$w = \frac{\Gamma(p+q+3)}{\Gamma(p+2)} (1 - [u_2\, \Gamma(q+2)]^{1/(q+1)}).$$

Consequently, after some rearranging, this becomes

$$\Pr[s + t \leqslant 1] = \frac{\Gamma(p+q+3)}{\Gamma(q+1)\,\Gamma(p+2)} \int_0^{[\Gamma(q+2)]^{1/(q+1)}} z^q (1-z)^{p+1}\, dz \tag{4.16}$$

The integral in (4.16) is essentially the density function for the beta distribution with parameters q and $p + 1$. It has been extensively tabulated as the incomplete beta distribution [88, pp. 142–156].

To generate the F distribution with m and n degrees of freedom, then $p = (m/2) - 1$ and $q = (n/2) - 1$, so that (4.12) becomes

$$\Pr[s+t \leqslant 1] = \frac{\Gamma((m+n+2)/2)}{\Gamma(n/2)\Gamma((m+2)/2)} \int_0^{[\Gamma((n+2)/2)]^{2n}} z^{(n-2)/2} \cdot (1-z)^{m/2}\, dz$$

$$= \frac{(m+n)\Gamma((m+n)/2)}{\Gamma(m)(m/2)\Gamma(n/2)} \int_0^{[(n/2)\Gamma(n/2)]^{2/n}} z^{(n-2)/2} (1-z)^{m/2}\, dz, \tag{4.17}$$

which is associated with the beta distribution with parameters $(n/2) - 1$ and $m/2$.

The Student's t distribution, with n degrees of freedom, is obtained from the F distribution, with one and n degrees of freedom. Consequently, $m = 1$ in (4.17), and

$$\Pr[s+t \leqslant 1] = \frac{(n+1)\Gamma(n+1)/2}{\sqrt{\pi}\,\Gamma(n/2)} \int_0^{[(n/2)\Gamma(n/2)]^{2/n}} z^{(n-2)/2} (1-z)^{1/2} dz \tag{4.18}$$

for the Student's t distribution. Here (4.18) is associated with the beta distribution, with parameters $(n/2) - 1$ and $\frac{1}{2}$, so that a published table could be consulted for this case as well as the preceding two.

In certain cases, at least one of the parameters in the incomplete beta integral is an integer, and this can simplify to an integral of the form

$$\Pr[s + t \leqslant 1] = \frac{\Gamma(p+q+2)}{\Gamma(p+1)\Gamma(q+1)} \int_0^x t^p (1-t)^q dt. \tag{4.19}$$

When q is a positive integer, it is helpful to note that

$$(1-t)^q = \sum_{k=0}^{q} \binom{q}{k} (-t)^k,$$

so that (4.19) can be written as the finite series

$$\Pr[s + t \leqslant 1] = \frac{\Gamma(p+q+2)}{\Gamma(p+1)q!} x^{p+1} \sum_{k=0}^{q} \binom{q}{k} \frac{(-x)^k}{p+1+k}$$

If it occurs that p is a positive integer, repeated integration by parts will yield the following finite series:

$$\Pr[s + t \leqslant 1] = 1 - \sum_{k=0}^{p} \left\{ p + \binom{q}{k} + 1 \right\} x^k (1-x)^{p+q+1-k}.$$

If, however, both p and q are positive integers, then repeated integration by parts can be used to find that

$$\Pr[s + t \leqslant 1] = \sum_{k=p+1}^{p+q+1} \left\{ p + \binom{q}{k} + 1 \right\} x^k (1-x)^{p+q+1+k}.$$

9 The Poisson distribution

If the random variable K has the Poisson distribution, then K has density function

$$f(k) = e^{-\lambda} \frac{\lambda^k}{k!} \tag{4.20}$$

where $\lambda > 0$ and k is a non-negative integer. Here, then,

$$f(k) = \Pr[K = k].$$

Recall that if $X_1, X_2, \ldots, X_k$ are independent and each distributed as an exponential with parameter λ, then

$$Y_k = \sum_{i=1}^{k} X_i ,$$

has the Gamma distribution $\Gamma(\lambda, k)$. It is interesting to note the distribution of K, where K is defined implicitly by the inequalities

$$Y_K = \sum_{i=1}^{K} X_i \leqslant 1 < \sum_{i=1}^{K+1} X_i . \tag{4.21}$$

If H is the distribution function of K, then

$$H(k) = \Pr[K \leqslant k],$$

and the density function of K is given by

$$h(k) = H(k) - H(k-1).$$

Now recalling (4.9), we have

$$\Pr[K \geqslant k] = \int_0^1 \frac{\lambda(\lambda y)^{k-1}}{(k-1)!} e^{-\lambda y}\, dy$$

$$= 1 - H(k-1).$$

Thus

$$H(k) = \int_1^\infty \frac{\lambda(\lambda y)^k}{k!} e^{-\lambda y}\, dy, \text{ and}$$

$$h(k) = \int_1^\infty \left(\frac{\lambda^{k+1} y^k e^{-\lambda y}}{k!} - \frac{\lambda^k y^{k-1} e^{-\lambda y}}{(k-1)!} \right) dy$$

$$= e^{-\lambda} \frac{\lambda^k}{k!},$$

so that K has the Poisson distribution.

10 The binomial distribution

If the random variable X has the binomial distribution, then X has density function [76, p. 64]

$$f(x) = \binom{n}{x} p^x (1-p)^{n-x},$$

where $x = 0, 1, \ldots, n$ and $0 < p < 1$. The distribution function of X is

$$F(x) = \sum_{t=0}^{x} f(t) = \sum_{t=0}^{x} \binom{n}{t} p^t (1-p)^{n-t}$$

$$= 1 - \sum_{t=x+1}^{n} \binom{n}{t} p^t (1-p)^{n-t}$$

where we note that $F(x) = P[X \leqslant x]$, and $f(x) = P[X = x]$.

Since X can take only a finite number of values, it is feasible to evaluate the distribution function F for each admissible value, and for each value of X, $0 \leqslant F(x) \leqslant 1$. This suggests a simple technique for the generation of pseudo-random numbers from the binomial distribution. Indeed, this method can be used to generate any random variable assuming only a finite number of values.

Let U denote a random variable having the uniform distribution over [0, 1]. Then for every u, there exists a unique value of x such that

$$F(x - 1) \leqslant u < F(x),$$

so that letting $X = x$ will provide the desired pseudo-random number.

Chapter 5

THE WISHART AND THE MULTIVARIATE NORMAL DISTRIBUTIONS

1 The multivariate normal distribution

In the following the term "n-vector" shall be interpreted as an $n \times 1$ matrix of real numbers. We write $\mathbf{X} \sim N(\boldsymbol{\mu}, \boldsymbol{\Sigma})$ to indicate that the random n-vector $\mathbf{X}$ satisfies a multivariate normal distribution with mean vector $\boldsymbol{\mu}$ and covariance matrix $\boldsymbol{\Sigma}$. Here

$$\boldsymbol{\mu} = \mathrm{E}(\mathbf{X}) \quad \text{and} \quad \boldsymbol{\Sigma} = \mathrm{E}[\mathbf{X} - \boldsymbol{\mu})(\mathbf{X} - \boldsymbol{\mu})^{\mathrm{T}}] \qquad (5.1)$$

where E denotes the expectation operator and the superscript T denotes matrix transposition. The probability density function of $\mathbf{X}$ is given by

$$f(\mathbf{x}) = ((2\pi)^p \; |\boldsymbol{\Sigma}|)^{-1/2} \exp\{-\tfrac{1}{2} (\mathbf{x} - \boldsymbol{\mu})^{\mathrm{T}} \boldsymbol{\Sigma}^{-1} (\mathbf{x} - \boldsymbol{\mu})\}. \qquad (5.2)$$

For a general exposition of multivariate normal statistics the reader is referred to the book on the subject by T. W. Anderson[1]. In particular, we cite the following result without proof.

Theorem 5.1 If the n-vector $\mathbf{X}$ is distributed $N(\boldsymbol{\mu}, \boldsymbol{\Sigma})$, $\boldsymbol{\tau}$ is a fixed m-vector, and $\mathbf{A}$ is an $m \times n$ matrix of rank m, then $\mathbf{Y} = \mathbf{AX} + \boldsymbol{\tau}$ is distributed $N(\mathbf{A}\boldsymbol{\mu} + \boldsymbol{\tau}, \mathbf{A}\boldsymbol{\Sigma}\mathbf{A}^{\mathrm{T}})$.

The above theorem provides a convenient means of generating a random vector $\mathbf{X}$ with any specified $\boldsymbol{\mu}$ and covariance $\boldsymbol{\Sigma}$, provided we can meet two basic requirements:

(i) we have a means of generating a random vector $\mathbf{Y}$ with mean $\mathbf{0}$ and covariance $\boldsymbol{I}$ where $\boldsymbol{I}$ is the identity matrix; and

(ii) we have a convenient means of factoring $\boldsymbol{\Sigma}$ in the form $\boldsymbol{\Sigma} = \mathbf{AA}^{\mathrm{T}}$ where $\mathbf{A}$ is an $n \times n$ matrix.

If the above two requirements are met, then $\mathbf{X} = \mathbf{AY} + \boldsymbol{\mu}$ has the desired distribution.

Concerning the first requirement, it is easily seen that if

$X_1, X_2, \ldots, X_n$ are one-dimensional random variables, independent, and all being distributed $N(0, 1)$, then the random vector $\mathbf{X} = (X_1, X_2, \ldots, X_n)^{\mathrm{T}}$ is distributed $N(\mathbf{0}, \boldsymbol{I})$. Let us then assume that we have at our disposal several random number generators operating independently and producing standardized normal variates.

For the second requirement there are several possibilities. Since $\boldsymbol{\Sigma}$ is positive definite (by assumption), there is a matrix $\boldsymbol{B}$ such that $\boldsymbol{B}\,\boldsymbol{\Sigma}\,\boldsymbol{B}^{\mathrm{T}} = \boldsymbol{I}$. In this event, $\boldsymbol{B}$ is non-singular and $\boldsymbol{\Sigma} = \boldsymbol{B}^{-1}(\boldsymbol{B}^{-1})^{\mathrm{T}}$. Thus it is always possible to satisfy the second condition. However, it is seldom practical to carry out the computations needed to find the matrix $\boldsymbol{B}^{-1}$ above, as this is tantamount to finding an orthogonal set of eigenvectors of $\boldsymbol{\Sigma}$.

A much more attractive solution of the second stipulation is furnished by the so-called Crout factorization of $\boldsymbol{\Sigma} = \{\sigma_{ij}\}$.

Theorem 5.2 Let Σ be a positive definite real symmetric matrix. There exists a lower triangular matrix $\mathbf{A}$ with positive elements on the main diagonal such that $\boldsymbol{\Sigma} = \mathbf{A}\mathbf{A}^{\mathrm{T}}$.

Theorem 5.2 is proved in [34] and the proof will not be given here. Note, however, that the elements of $\mathbf{A}$ can be computed recursively in the order 11, 21, 31, . . . , $n1$, 22, . . . , nn, that is, by columns. To see this, note that the condition of $\mathbf{A}$ gives $a_{ij} = 0$ for $j > i$. Hence we have

$$\sigma_{ij} = \sum_{k=1}^{j} a_{ik}\, a_{jk}\,. \tag{5.3}$$

For $i = j = 1$ we have $\sigma_{11} = a_{11}^2$ so that

$$a_{11} = \sigma_{11}^{1/2}\,. \tag{5.4}$$

The remaining elements in the first column of $\mathbf{A}$ are then given by

$$a_{i1} = \sigma_{i1} / a_{11}. \tag{5.5}$$

Once we have computed the first $j - 1$ columns of $\mathbf{A}$, we have

$$a_{jj} = \left(\sigma_{jj} - \sum_{k=1}^{j-1} a_{jk}^2\right)^{1/2} \tag{5.6}$$

Now if $j = n$, our task is completed. Otherwise we have

$$a_{ij} = \left(\sigma_{ij} - \sum_{k=1}^{j-1} a_{ik}\, a_{jk} / a_{jj}\right) \tag{5.7}$$

for $i = j + 1, \ldots, n$.

As an illustration of the preceding factorization, suppose that

$$\boldsymbol{\Sigma} = \begin{bmatrix} 4 & 2 & 2 \\ 2 & 5 & 1 \\ 1 & 1 & 5 \end{bmatrix}$$

Using (5.4) we have $a_{11} = 2$, and then using (5.5) we obtain $a_{21} = 1$, $a_{31} = 1$. Now by (5.6), $a_{22} = (5 - 1)^{1/2} = 2$. Then by (5.7), we have $a_{32} = (1 - 1)/2 = 0$. Finally, (5.6) gives $a_{33} = (5 - 1 - 0)^{1/2} = 2$. Thus,

$$\mathbf{A} = \begin{bmatrix} 2 & 0 & 0 \\ 1 & 2 & 0 \\ 1 & 0 & 2 \end{bmatrix}$$

The method outlined above seems to be a superior way to generate random vectors with any specified normal distribution. Other methods are available for generating normal random vectors, but differ from the above only in the factorization of $\boldsymbol{\Sigma}$. Of course, if a factorization $\boldsymbol{\Sigma} = BB^{\mathrm{T}}$ presents itself in a natural way, other than that given in Theorem 4.2, then the transformation $\mathbf{Z} = B\mathbf{X} + \mu$ will result in $\mathbf{Z} \sim N(\mu, \boldsymbol{\Sigma})$, provided that $\mathbf{X} \sim N(0, I)$.

2 The Wishart distribution

Suppose that $\mathbf{X}^{(1)}, \mathbf{X}^{(2)}, \ldots, \mathbf{X}^{(n)}$, $n \geq k$ are independent random k-vectors, each being distributed $N(\mu, \boldsymbol{\Sigma})$. Then the random matrix

$$\mathbf{V} = \sum_{i=1}^{n} (\mathbf{X}^{(i)} - \mu)(\mathbf{X}^{(i)} - \mu)^{\mathrm{T}} \tag{5.8}$$

is distributed according to the Wishart distribution $W(k, n, \boldsymbol{\Sigma})$. Note that $\mathbf{V}$ is a random k^2-vector, that is, a $k \times k$ matrix. The probability density function of $\mathbf{V} = (V_{ij})$ is given in [1] and [110]. Since the expression is lengthy, and its form is not important here, we shall omit it. $\mathbf{V}$ is called the "sample scatter about the population mean."

Let us observe that for purposes of simulating values of $\mathbf{V}$ it will suffice to assume that we are dealing only with the case $\boldsymbol{\mu} = 0$, $\boldsymbol{\Sigma} = \boldsymbol{I}$. This is seen immediately from the following result:

Theorem 5.3 Let $\boldsymbol{\Sigma} = \mathbf{A}\mathbf{A}^{\mathrm{T}}$ and suppose that $\mathbf{X}^{(i)} \sim N(\mathbf{0}, \boldsymbol{I})$, $i = 1, 2, \ldots, n$. Let

$$\mathbf{V}_x = \sum_{i=1}^{n} \mathbf{X}^{(i)} \mathbf{X}^{(i)\mathrm{T}}, \quad \mathbf{Y}^{(i)} = \mathbf{A}\mathbf{X}^{(i)} + \boldsymbol{\mu},$$

where $\mathbf{A}$ is $p \times k$ of rank $p \leqslant k$, and

$$\mathbf{V}_y = \sum_{i=1}^{n} (\mathbf{Y}^{(i)} - \boldsymbol{\mu})(\mathbf{Y}^{(i)} - \boldsymbol{\mu})^{\mathrm{T}}.$$

Then

$$\mathbf{V}_y = \mathbf{A}\mathbf{V}_x \mathbf{A}^{\mathrm{T}}. \tag{5.9}$$

Proof Suppressing limits on the summation symbol, we have

$$\begin{aligned}
\mathbf{V}_y &= \Sigma (\mathbf{Y}^{(i)} - \boldsymbol{\mu})(\mathbf{Y}^{(i)} - \boldsymbol{\mu})^{\mathrm{T}} \\
&= \Sigma (\mathbf{A}\mathbf{X}^{(i)} + \boldsymbol{\mu} - \boldsymbol{\mu})(\mathbf{A}\mathbf{X}^{(i)} + \boldsymbol{\mu} - \boldsymbol{\mu})^{\mathrm{T}} \\
&= \Sigma\, \mathbf{A}\mathbf{X}^{(i)}\mathbf{X}^{(i)\mathrm{T}}\mathbf{A}^{\mathrm{T}} \\
&= \mathbf{A}(\Sigma\, \mathbf{X}^{(i)}\mathbf{X}^{(i)\mathrm{T}})\mathbf{A}^{\mathrm{T}} \\
&= \mathbf{A}\mathbf{V}_x\mathbf{A}^{\mathrm{T}}.
\end{aligned}$$

Corollary 5.4 If $\mathbf{V} \sim W(k, n, \boldsymbol{I})$, then for any $p \times k$ matrix $\mathbf{A}$ of rank $p \leqslant k$ we have $\mathbf{A}\mathbf{V}\mathbf{A}^{\mathrm{T}} \sim W(p, n, \mathbf{A}\mathbf{A}^{\mathrm{T}})$.

Probably the most direct means of simulating a sample from the Wishart distribution $W(k, n, \boldsymbol{I})$ would be to generate a sample of pn independent variates from $N(0, 1)$ which we arrange in a $k \times n$ matrix

$$\boldsymbol{B} = \begin{bmatrix} b_{11} & b_{12} & \cdots & b_{1n} \\ b_{21} & b_{22} & \cdots & b_{2n} \\ \cdot & & & \\ \cdot & & & \\ \cdot & & & \\ b_{k1} & b_{k2} & \cdots & b_{kn} \end{bmatrix} \tag{5.10}$$

Using the columns $\mathbf{B}^{(1)}, \mathbf{B}^{(2)}, \ldots, \mathbf{B}^{(n)}$ of B for a sample of n vectors from the k-dimensional distribution $N(\mathbf{0}, I)$, we can calculate

$$\mathbf{V} = \sum_{i=1}^{n} \mathbf{B}^{(i)} \mathbf{B}^{(i)^{\mathrm{T}}} .$$

Clearly this is the result of forming the matrix product BB^{T}. That is,

$$V = BB^{\mathrm{T}} . \tag{5.11}$$

From the above we will derive a different method of generating V, which, if not more expedient, will present $\mathbf{V}$ in factored form $V = \mathbf{A}\mathbf{A}^{\mathrm{T}}$, where $\mathbf{A}$ is a $k \times k$ lower triangular matrix with positive elements on the diagonal. A secondary advantage of this method is that it makes the determinant of $\mathbf{V}$ readily available, since $|\mathbf{V}| = |\mathbf{A}\mathbf{A}^{\mathrm{T}}| = |\mathbf{A}|^2 = (a_{11}a_{12} \ldots a_{nn})^2$. Although the expression for the probability density function for $\mathbf{V}$ has been omitted, we at least mention that $|\mathbf{V}|$ is needed for the evaluation of this function.

Using B from (5.10), let $\mathbf{B}_i$, $1 \leqslant i \leqslant k$, be the ith row of B and note that the $\mathbf{B}_i^{\mathrm{T}}$ are independent n-vectors distributed $N(\mathbf{0}, I)$. By applying the Gram–Schmidt orthogonalization process to the rows of B, we can define row vectors $\mathbf{Y}_i$, $1 \leqslant i \leqslant k$, given recursively by

$$\mathbf{Y}_j = \mathbf{B}_j (I - \mathbf{Q}_1 - \mathbf{Q}_2 - \ldots - \mathbf{Q}_{j-1}), \tag{5.12}$$

where $\mathbf{Q}_r$ is the $n \times n$ matrix

$$\mathbf{Q}_r = \mathbf{Y}_r^{\mathrm{T}} \mathbf{Y}_r / \mathbf{Y}_r \mathbf{Y}_r^{\mathrm{T}} . \tag{5.13}$$

From the definition of $\mathbf{Y}_j$ it is clear that, for $i > j$, the quantity $\mathbf{Y}_j$ is independently distributed with respect to $\mathbf{B}_i$. We conclude immediately that

$$A_{ij} = (\mathbf{B}_i \mathbf{Y}_i^{\mathrm{T}}) / \sqrt{\mathbf{Y}_j \mathbf{Y}_j^{\mathrm{T}}}, \quad i > j \tag{5.14}$$

is distributed $N(0, 1)$.

To proceed further, we need the results of the well-known Cochran–Fisher theorem [110].

Theorem 5.5 If $\mathbf{X}$ is a random n-vector and is distributed $N(\mathbf{0}, I)$, and if

$$\mathbf{X}^{\mathrm{T}}\mathbf{X} = \sum_{j=1}^{k} \mathbf{X}^{\mathrm{T}} \mathbf{Q}_j \mathbf{X} \tag{5.15}$$

where each $\mathbf{Q}_j$ is positive semidefinite of rank n_j, then a necessary and sufficient condition for the $\mathbf{X}^T \mathbf{Q}_j \mathbf{X}$ to be independently distributed as a chi-square with n_j degrees of freedom, $j = 1, 2, \ldots, k$, is that $n = n_1 + n_2 + \ldots + n_k$.

The preceding theorem is applied as follows. Let i be fixed, $1 \leqslant i \leqslant k$ and let

$$\mathbf{Q} = I - \mathbf{Q}_1 - \mathbf{Q}_2 - \ldots - \mathbf{Q}_{i-1}. \tag{5.16}$$

Since each $\mathbf{Q}_j$ is of rank 1 and they satisfy $\mathbf{Q}_j \mathbf{Q}_l = 0$ if $j \neq l$, it follows that the rank of $\mathbf{Q}$ is $n - i + 1$. Now since $\mathbf{B}_i$ is independent of $\mathbf{Q}_1, \mathbf{Q}_2, \ldots, \mathbf{Q}_{i-1}$, it is also independent of $\mathbf{Q}$. Using the definition of Q, we have

$$\mathbf{B}_i \mathbf{B}_i^T = \mathbf{B}_i \mathbf{Q} \mathbf{B}_i^T + \sum_{j=1}^{i-1} \mathbf{B}_i \mathbf{Q}_j \mathbf{B}_i^T. \tag{5.17}$$

From (5.17) we can draw several conclusions by applying Theorem 5.5:

(a) $\mathbf{Y}_i \mathbf{Y}_i^T = \mathbf{B}_i \mathbf{Q} \mathbf{B}_i^T \sim \chi^2 (n - i + 1)$.

Note that $\mathbf{B}_i \mathbf{Q} \mathbf{B}_i^T = \mathbf{Y}_i \mathbf{B}_i^T = \mathbf{B}_i \mathbf{Y}_i^T = \mathbf{Y}_i \mathbf{Y}_i^T$

(b) $\mathbf{B}_i \mathbf{Q}_j \mathbf{B}_i^T \sim \chi^2 (1)$,

a fact that we know since $\mathbf{B}_i \mathbf{Q}_j \mathbf{B}_i^T = (\mathbf{B}_i \mathbf{Y}_i^T / \sqrt{\mathbf{Y}_j \mathbf{Y}_j^T})^2$ and is thus the square of a $N(0, 1)$ variate (see 5.14).

(c) $\mathbf{B}_i \mathbf{Q} \mathbf{B}_i^T$ and $\mathbf{B}_i \mathbf{Q}_j \mathbf{B}_i^T$, $i > j$ are independent.

Now from (5.12) and (5.13) it is easily seen that the conditional random variable $(\mathbf{Y}_i \mid \mathbf{Y}_1, \mathbf{Y}_2, \ldots, \mathbf{Y}_{i-1})$ is normally distributed. Since the $\mathbf{Y}_i$ are mutually orthogonal it follows that the quantities $\mathbf{Y}_i \mathbf{Y}_i^T$ are independently distributed and, from property (a) above, as $\chi^2 (n - i + 1)$.

Let us define

$$A_{ii} = \sqrt{\mathbf{Y}_i \mathbf{Y}_i^T} \qquad A_{ij} = 0, \quad i < j, \tag{5.18}$$

and note that from (5.14), for $i > j$,

$$A_{ij} = \mathbf{B}_i \mathbf{Y}_i^T / A_{jj}. \tag{5.19}$$

It remains to observe that for $i > j$, $i' > j'$ and $j \neq j'$

the quantities A_{ij} and $A_{i'j'}$ are independent, since $\mathbf{B}_i$ and $\mathbf{B}_{i'}$ are independent.

Now $\mathbf{A} = (A_{ij})$ is a lower triangular matrix. Letting C_{ij} be the ij element of $\mathbf{AA}^{\mathrm{T}}$, we have

$$C_{ii} = A_{ii}^2 = \mathbf{Y}_i \mathbf{Y}_i^{\mathrm{T}} = \mathbf{B}_i \mathbf{B}_i^{\mathrm{T}}$$

and for $i < l$, using the facts that $A_{ij} = 0$ for $i < j$, and that $\mathbf{B}_i \mathbf{Y}_j^{\mathrm{T}} = \mathbf{Y}_j \mathbf{B}_i^{\mathrm{T}}$, we have

$$C_{il} = \sum_{j=1}^{i} A_{ij} A_{lj}$$

$$= \sum_{j=1}^{i} (\mathbf{B}_i \mathbf{Y}_j^{\mathrm{T}}/A_{jj})(\mathbf{Y}_j \mathbf{B}_l^{\mathrm{T}}/A_{jj})$$

$$= \sum_{j=1}^{i} (\mathbf{B}_i \mathbf{Y}_j^{\mathrm{T}} \mathbf{Y}_j \mathbf{B}_l^{\mathrm{T}})/A_{jj}^2$$

$$= \mathbf{B}_i \left(\sum_{j=1}^{i} \mathbf{Y}_j^{\mathrm{T}} \mathbf{Y}_j / \mathbf{Y}_j \mathbf{Y}_j^{\mathrm{T}} \right) \mathbf{B}_l^{\mathrm{T}}$$

$$= \mathbf{B}_i \mathbf{B}_l^{\mathrm{T}} .$$

Now, the preceding shows that $\mathbf{AA}^{\mathrm{T}} = \boldsymbol{BB}^{\mathrm{T}} = \mathbf{V}$. We may summarize these results in the following.

Theorem 5.6 If the random matrix $\mathbf{V}$ satisfies the Wishart distribution $W(k, n, \boldsymbol{I})$, and if $\mathbf{A}$ is the lower triangular matrix with positive elements on the diagonal such that $\mathbf{V} = \mathbf{AA}^{\mathrm{T}}$, then the non-zero elements A_{ij} of $\mathbf{A}$ are independently distributed where A_{ii} is the square root of a random variable distributed as $\chi^2(n - i + 1)$ and A_{ij}, $i > j$, is $N(0, 1)$.

These results were essentially due to Odell and Feiveson [83] in a slightly different form. Note that by the unicity of the Crout factorization the converse of Theorem 5.6 also holds, namely:

Corollary 5.7 If $\mathbf{A}$ is as above, then $\mathbf{AA}^{\mathrm{T}} \sim W(k, n, \boldsymbol{I})$.

More generally, if $\mathbf{A}$ is as above and if $\boldsymbol{B}$ is a $p \times k$ matrix of rank $p \leqslant k$, then $\boldsymbol{BAA}^{\mathrm{T}}\boldsymbol{B}^{\mathrm{T}} \sim W(p, n, \boldsymbol{\Sigma})$, where $\boldsymbol{\Sigma} = \boldsymbol{BB}^{\mathrm{T}}$.

The second part of Corollary 5.7 follows immediately from the first part by means of Corollary 5.5.

Note that if $\boldsymbol{B}$ is $k \times k$ and is the Crout factorization of a $k \times k$ covariance matrix $\boldsymbol{\Sigma}$, then $\boldsymbol{BA}$ is the Crout factorization of $\boldsymbol{V} = \boldsymbol{BAA}^{\mathrm{T}}\boldsymbol{B}^{\mathrm{T}}$.

As compared with the brute force method of generating Wishart variates suggested following Corollary 5.4, which requires $n \times k$ random numbers, the above method is apparently simpler, requiring only $k(k+1)/2$ random numbers. One should note that the diagonal elements V_{ii} are easily obtained by means of the Wilson–Hilferty χ^2 approximation (see Chapter 4).

3 Sample covariance

If in (5.8) we replace the population mean by the sample mean

$$\bar{\mathbf{X}} = \frac{1}{n} \sum_{i=1}^{n} \mathbf{X}^{(i)}$$

and assume that $n > k$, then

$$\boldsymbol{V} = \sum_{i=1}^{n} (\mathbf{X}^{(i)} - \bar{\mathbf{X}})(\mathbf{X}^{(i)} - \bar{\mathbf{X}})^{\mathrm{T}} \tag{5.20}$$

satisfies $W(k, n-1, \boldsymbol{\Sigma})$ (see [1]). Here we are assuming as before that the k-vectors $\mathbf{X}^{(i)}$ are identically distributed $N(\mu, \boldsymbol{\Sigma})$ and are independent. It is well known that the sample covariance

$$\boldsymbol{S} = \frac{1}{n} \boldsymbol{V} \tag{5.21}$$

is the maximum likelihood estimator of $\boldsymbol{\Sigma}$.

In testing point-clustering techniques related to pattern recognition problems it is necessary to simulate values of $\boldsymbol{S}$. Using Corollary 5.7, values of $\boldsymbol{S}$ may be generated by generating values of $\boldsymbol{A}$ as given in Theorem 5.6. Then $\boldsymbol{S} = \frac{1}{n}\boldsymbol{BAA}^{\mathrm{T}}\boldsymbol{B}^{\mathrm{T}}$, $\boldsymbol{BB}^{\mathrm{T}} = \boldsymbol{\Sigma}$. If $\boldsymbol{B}$ is the Crout factorization of $\boldsymbol{\Sigma}$, then $(\boldsymbol{BA})/\sqrt{n}$ is the Crout factorization of $\boldsymbol{S}$; a feature which is extremely desirable in pattern recognition simulation.

Chapter 6

SIMULATION OF NORMAL PROCESSES

1 Introductory concepts [32]

A *random process* or *time series* is a collection of random variables $X(t)$, one for each t in some index set. A random process $X(t)$ is said to be normal (or Gaussian) provided that for any indices $t_1, t_2, \ldots, t_n$, the random variables $X(t_1), X(t_2), \ldots, X(t_n)$ have a multivariate normal distribution. In most cases it happens that t assumes only real-number values, and we so suppose in all that follows. Moreover, we restrict our attention to normal processes. A process $X(t)$ is said to be strictly stationary provided that $t_1 - s_1 = t_2 - s_2 = \ldots = t_n - s_n$ implies that $(X(t_1), \ldots, X(t_n))$ and $(X(s_1), \ldots, X(s_n))$ have the same joint distribution fot any integer n.

In dealing with normal processes it is convenient to make the simplifying assumption that $\mathrm{E}(X(t)) = 0$, for all t, where E is the expectation operator. Otherwise we replace $X(t)$ with the normal process $X(t) - \mathrm{E}(X(t))$. On this assumption, the autocorrelation function $R(t, s)$ is defined by

$$R(t, s) = \mathrm{E}(X(t)\, X(s)). \qquad (6.1)$$

In the case where $X(t)$ is covariance stationary [86], $R(t, s)$ depends only on the difference $t - s = \tau$ and we write

$$R(\tau) = \mathrm{E}(X(t)X(t - \tau)) \qquad (6.2)$$

It is a fact that a normal process that is covariance stationary is strictly stationary. The Fourier transform [86]

$$S(\omega) = \int_{-\infty}^{\infty} R(\tau)e^{-i\omega\tau}\, d\tau \qquad (6.3)$$

is useful in studying stationary processes, and it is shown in [86] and elsewhere that $S(\omega)$ determines the density for the distribution of power among the various frequencies. Thus, $S(\omega)$ is referred to as the "power spectral density" [86] of the process. In general, the Fourier transform (6.3) can be inverted so that $S(\omega)$ also determines $R(\tau)$. At times it is more natural to work with $S(\omega)$ rather than $R(\tau)$. One must be cautious, however, since small perturbations in $S(\omega)$ can cause large variations in $R(\tau)$.

2 Simulation of normal random processes

The method presented here applies equally well to stationary and non-stationary processes. In the case of the former there is an alternative method which has certain advantages and will be presented in section 3, following.

Assume that we wish to sample the normal process $S(t)$ at the points $t_1, t_2, \ldots, t_n$. Since $X(t_1), \ldots, X(t_n)$ are jointly distributed as multivariate normal, we need only determine the covariance matrix $\mathbf{\Sigma} = \mathbf{\Sigma}(t_1, t_2, \ldots, t_n)$ and then sample the distribution $N(\mathbf{0}, \mathbf{\Sigma})$ by the methods of Chapter 5.

Now let $\mathbf{\Sigma} = (\sigma_{ij})$ and note that

$$\sigma_{ij} = \mathrm{E}(X(t_i)\, X(t_j)) = R(t_i, t_j) \tag{6.4}$$

for $1 \leqslant i, j \leqslant n$. Thus $\mathbf{\Sigma}$ is found by evaluating $R(t, s)$ at the pairs of sampling points.

Although we tacitly assumed in Chapter 5 that $\mathbf{\Sigma}$ was positive definite, we have no reason to expect this in the present context. However, if $\mathbf{\Sigma}$ has rank $k < n$, then there are k points which we may assume to be $t_1, t_2, \ldots, t_k$ (permuting the indices if necessary) such that the conditional distribution of $X(t_{k+1}), \ldots, X(t_n)$, given $X(t_1), \ldots, X(t_k)$, is degenerate. Hence we may restrict our attention to $X(t_1), \ldots, X(t_k)$, which have a positive definite covariance, and can calculate the remaining values in terms of these.

Note that the method is quite general and involves only knowing the autocorrelation function of the process.

3 Simulation of stationary normal processes

It is shown in [21] that if the stationary processes $X(t)$ and $W(t)$ are related by means of the linear transformation

$$X(t) = \int_{-\infty}^{t} g(t-s)\, W(s)\, ds, \tag{6.5}$$

then the respective power spectral densities $S_X(\omega)$ and $S_W(\omega)$ are related by

$$S_X(\omega) = |G(\omega)|^2 S_W(\omega), \tag{6.6}$$

where $G(\omega)$ is the Fourier transform of $g(t)$.

Equation (6.6) is the basis for a method due to J.N. Franklin [32] for sampling the stationary process $X(t)$. The idea is to choose a function $G(\omega)$ satisfying

$$S_X(\omega) = |G(\omega)|^2. \tag{6.7}$$

One then chooses a process $W(t)$ with $S_W(\omega) \equiv 1$. Such a process is an idealized process known as "white noise". Note that one would necessarily have $R_W(\tau) = \delta(\tau)$, and thus $W(t)$ has infinite variance for all t. The symbol $\delta(\tau)$ denotes the Dirac delta function [21].

Now if we assume that $S(\omega) = S_X(\omega)$ is a rational function and satisfies the conditions

$$0 \leqslant S(\omega) < \infty, \quad S(\omega) = S(-\omega), \quad S_\omega \to 0 \quad \text{as } \omega \to \infty,$$

then it can be shown [21] that $S(\omega)$ can be represented in the form

$$S(\omega) = \left|\frac{P(i\omega)}{Q(i\omega)}\right|^2, \quad \omega \text{ real}, \tag{6.8}$$

where $P(z)$ and $Q(z)$ are polynomials with real coefficients, the degree of P is less than that of Q, and the zeros of $Q(z)$ lie in the half-plane Re $z < 0$. Then equation (6.7) can be satisfied by taking

$$G(\omega) = \frac{P(i\omega)}{Q(i\omega)}. \tag{6.9}$$

Letting $D = d/dt$ and using the properties of a Fourier transform, it is easy to see that an equivalent formulation of (6.5) is given by

$$X(t) = [P(D)/Q(D)]\, W(t). \tag{6.10}$$

In the last equation we first find the steady-state solution $\phi(t)$ of

$$Q(D)\,\phi(t) = W(t) \tag{6.11}$$

and calculate $X(t)$ as

$$X(t) = P(D)\,\phi(t). \tag{6.12}$$

Now assume that (6.11) has the form

$$\phi^{(n)}(t) + a_1 \phi^{(n-1)}(t) + \ldots + a_n \phi(t) = W(t). \qquad (6.13)$$

We wish to compute samples of $X(t)$ at the points 0, Δt, $2\Delta t, \ldots$, so that we need samples of the state vector

$$\mathbf{Z}(t) = [\phi(t), \phi^{(1)}(t), \ldots, \phi^{(n-1)}(t)]^{\mathrm{T}} \qquad (6.14)$$

at these points. Now the vector $\mathbf{Z}(t)$ is seen to satisfy the stochastic differential equation

$$\frac{d}{dt}\mathbf{Z}(t) = \mathbf{A}\mathbf{Z}(t) + \mathbf{f}(t), \quad -\infty < t < \infty, \qquad (6.15)$$

where

$$\mathbf{A} = \begin{bmatrix} 0 & 1 & 0 & \ldots & 0 \\ 0 & 0 & 1 & \ldots & 0 \\ \vdots & \vdots & \vdots & & \vdots \\ 0 & 0 & 0 & \ldots & 1 \\ -a_n & -a_{n-1} & -a_{n-2} & \ldots & -a_1 \end{bmatrix} \qquad (6.16)$$

and $\mathbf{f}(t) = [0,0,\ldots, 0, W(t)]^{\mathrm{T}}$.

We shall first compute $\mathbf{Z}(0)$, which is normal random vector with covariance matrix $\mathbf{\Sigma}$. Note that since $\mathbf{\Sigma}(t)$ is stationary

$$\mathbf{\Sigma} = \mathrm{E}(\mathbf{Z}(0)\,\mathbf{Z}(0)^{\mathrm{T}}) = \mathrm{E}(\mathbf{Z}(t)\,\mathbf{Z}(t)^{\mathrm{T}}) \qquad (6.17a)$$

Also, from the definition of $\mathbf{Z}(t)$, we have

$$\sigma_{ij} = \mathrm{E}(\phi^{(i-1)}\,\phi^{(j-1)}). \qquad (6.17b)$$

The matrix $\mathbf{\Sigma}$ can be computed by the following formulae, which are derived in [31] :

$$\sigma_{ij} = \begin{cases} 0 & \text{for } i+j \text{ odd} \\ (-1)^{(j-i)/2}\, m_{(i+j)/2} & \text{for } i+j \text{ even}, \end{cases} \qquad (6.18)$$

where the numbers $m_0, m_1, \ldots, m_{n-1}$ can be found by solving the equations

$$(-1)^k \sum_{k/2 \leqslant q \leqslant (n+k)/2} (-1)^q\, m_q\, a_{n-2q+k} = \begin{cases} 0, & k = 0, \ldots, n-2 \\ \frac{1}{2}, & k = n-1 \end{cases} \qquad (6.19)$$

For example, for $n = 6$, these equations are

$$\boldsymbol{\Sigma} = \begin{bmatrix} m_0 & 0 & -m_1 & 0 & m_2 & 0 \\ 0 & m_1 & 0 & -m_2 & 0 & m_3 \\ -m_1 & 0 & m_2 & 0 & -m_3 & 0 \\ 0 & -m_2 & 0 & m_3 & 0 & -m_4 \\ m_2 & 0 & -m_3 & 0 & m_4 & 0 \\ 0 & m_3 & 0 & -m_4 & 0 & m_5 \end{bmatrix} \tag{6.20a}$$

where

$$\begin{bmatrix} a_6 & -a_4 & a_2 & -1 & 0 & 0 \\ 0 & a_5 & -a_3 & a_1 & 0 & 0 \\ 0 & -a_6 & a_4 & -a_2 & 1 & 0 \\ 0 & 0 & -a_5 & a_3 & -a_1 & 0 \\ 0 & 0 & a_6 & -a_4 & a_2 & -1 \\ 0 & 0 & 0 & a_5 & -a_3 & a_1 \end{bmatrix} \begin{bmatrix} m_0 \\ m_1 \\ m_2 \\ m_3 \\ m_4 \\ m_5 \end{bmatrix} = \begin{bmatrix} 0 \\ 0 \\ 0 \\ 0 \\ 0 \\ \frac{1}{2} \end{bmatrix} \tag{6.20b}$$

Once $\boldsymbol{\Sigma}$ has been computed it can be factored by means of the Crout factorization (section 1 of Chapter 5) in the form $\boldsymbol{\Sigma} = BB^{\mathrm{T}}$, and $\mathbf{Z}(0)$ can be calculated by generating a vector $\mathbf{W}_0$ distributed $N(\mathbf{0}, \boldsymbol{I})$ and setting

$$\mathbf{Z}(0) = \boldsymbol{B}\mathbf{W}_0 . \tag{6.21}$$

It remains to show how one obtains $\mathbf{Z}(t + \Delta t)$ once $\mathbf{Z}(t)$ has been calculated. The matrix equation

$$\mathbf{A}\mathbf{C}(t) = D\mathbf{C}(t) = \frac{d}{dt}\mathbf{C}(t)$$

has the solution

$$\mathbf{C}(t) = (\exp \mathbf{A}t)\,\mathbf{C}(0).$$

Considered as an initial value problem with $\mathbf{C}(0) = \boldsymbol{I}$, the solution is then

$$\mathbf{C}(t) = \exp \mathbf{A}t .$$

We shall need to know the value of exp $\mathbf{A}t$ for the single value $t = \Delta t$. If the eigenvalues of $\mathbf{A}$ are known, then the computation of exp $\mathbf{A}\Delta t$ is easily accomplished by standard facts from matrix theory [34]. Otherwise, one can calculate it directly using other techniques, such as the Runge–Kutta method.

From the differential equation (6.15) for $\mathbf{Z}(t)$ we see that

$$\mathbf{Z}(t + \Delta t) = (\exp \mathbf{A}\Delta t)\, \mathbf{Z}(t) + \mathbf{R}(t), \tag{6.22}$$

where $\mathbf{R} = \mathbf{R}(t)$ is the normal random vector

$$\mathbf{R} = \int_0^{\Delta t} \exp((\Delta t - s)\mathbf{A}\, \mathbf{f}(t + s) ds. \tag{6.23}$$

Now observe that $\mathbf{Z}(t)$ can be written in the form

$$\mathbf{Z}(t) = \int_{-\infty}^{t} \exp(\mathbf{A}(t - s))\, \mathbf{f}(s)\, ds\,.$$

Moreover, for $s > 0$ and $s' < t$,

$$\mathrm{E}[\,\mathbf{f}(t + s)\,\mathbf{f}(s')^{\mathrm{T}}] = \mathbf{C}\delta(t + s - s') = \mathbf{0}\,,$$

where, from the definition of $\mathbf{f}(t)$,

$$\mathbf{C} = \begin{bmatrix} 0 & 0 & \cdots & 0 & 0 \\ 0 & 0 & & 0 & 0 \\ \cdot & & & \cdot & \cdot \\ \cdot & & & \cdot & \cdot \\ \cdot & & & \cdot & \cdot \\ 0 & 0 & & 0 & 0 \\ 0 & 0 & \cdots & 0 & 1 \end{bmatrix}$$

It follows immediately from these facts that $\mathbf{R}$ and $\mathbf{Z}(t)$ are uncorrelated. Thus, a sample of $\mathbf{R}$ can be generated independently of $\mathbf{Z}(t)$. To complete the determination of $\mathbf{Z}(t + \Delta t)$ it remains to determine the covariance matrix $\mathbf{\Sigma_R}$ of $\mathbf{R}$.

Now,

$$\begin{aligned} \mathbf{\Sigma_R} &= \mathrm{E}(\mathbf{R}\mathbf{R}^{\mathrm{T}}) \\ &= \mathrm{E} \int_0^{\Delta t}\!\int_0^{\Delta t} \exp[\mathbf{A}(\Delta t - s_1)]\, \mathbf{f}(t + s_1)\mathbf{f}(t + s_2)^{\mathrm{T}} \\ &\qquad\qquad \exp[\mathbf{A}^{\mathrm{T}}(\Delta t - s_2)]\, ds_1\, ds_2 \end{aligned}$$

$$= \int_0^{\Delta t}\int_0^{\Delta t} \exp[\mathbf{A}(\Delta t - s_1)]\,\mathbf{C}\delta(s_1 - s_2)\exp[\mathbf{A}^{\mathrm{T}}(\Delta t - s_2)]\,ds_1\,ds_2$$

$$= \int_0^{\Delta t} \exp[\mathbf{A}(\Delta t - s)]\,\mathbf{C}\exp[\mathbf{A}^{\mathrm{T}}(\Delta t - s)]ds$$

$$= \int_0^{\Delta t} \exp(\mathbf{A}s)\,\mathbf{C}\exp(\mathbf{A}^{\mathrm{T}}s)\,ds, \tag{6.24}$$

the last step being accomplished by a change of variable to $\Delta t - s$.

Let the integrand in the last expression of (6.24) be denoted by $\mathbf{J}(s)$ for convenience, and note that

$$\frac{d\mathbf{J}(s)}{ds} = \mathbf{A}\mathbf{J}(s) + \mathbf{J}(s)\mathbf{A}^{\mathrm{T}}.$$

Integration from 0 to Δt immediately gives the result

$$\exp(\mathbf{A}\Delta t)\,\mathbf{C}[\exp(\mathbf{A}\Delta t)]^{\mathrm{T}} - \mathbf{C} = \mathbf{A}\mathbf{\Sigma}_{\mathbf{R}} + \mathbf{\Sigma}_{\mathbf{R}}\mathbf{A}^{\mathrm{T}}. \tag{6.25}$$

Making use of the fact that the eigenvalues of $\mathbf{A}$ are in the half-plane Re $\mathbf{z} < 0$ it can be shown that this equation uniquely determines $\mathbf{\Sigma}_{\mathbf{R}}$. Observe that as a consequence of (6.25), $\mathbf{\Sigma}_{\mathbf{R}}$ does not depend on t, but only on Δt.

To determine $\mathbf{\Sigma}_{\mathbf{R}} = (\mu_{ij})$, let the elements of exp $\mathbf{A}\Delta t$ be denoted by e_{ij}. Written out, (6.25) becomes

$$\sum_{k=1}^{n} (a_{ik}\,\mu_{kj} + a_{jk}\,\mu_{ik}) = d_{ij} \tag{6.26}$$

where

$$d_{ij} = \begin{cases} e_{in}\,e_{jn} & \text{if } i < n \text{ or } j < n \\ e_{nn}^{2} - 1 & \text{if } i = j = n. \end{cases}$$

These n^2-equations involve only $n(n+1)/2$ unknowns, since $\mathbf{\Sigma}_{\mathbf{R}}$ is symmetric. In fact, since the equations in (6.26) are symmetric with respect to i and j it is enough to use only those equations for which $i \geqslant j$.

Once $\mathbf{\Sigma}_{\mathbf{R}}$ has been computed it can be factored by the Crout factorization into $\mathbf{\Sigma}_{\mathbf{R}} = \mathbf{B}_{\mathbf{R}}\,\mathbf{B}_{\mathbf{R}}^{\mathrm{T}}$. Now for $t + \Delta t = k\Delta t$, we generate $\mathbf{R}$ by generating a random vector $\mathbf{W}_k \sim N(\mathbf{0}, \mathbf{I})$ and set

$\mathbf{R} = \mathbf{B}\mathbf{W}_k$. Then $\mathbf{Z}(t + \Delta t)$ is given by

$$\mathbf{Z}(t + \Delta t) = \exp(\mathbf{A}\Delta t)\, \mathbf{Z}(t) + \mathbf{R},$$

or equivalently

$$\mathbf{Z}(k\Delta t) = \exp(\mathbf{A}\Delta t)\, \mathbf{Z}[(k - 1)\Delta t] + \mathbf{B}_{\mathbf{R}}\, \mathbf{W}_k. \qquad (6.27)$$

This process may be repeated indefinitely for $t = 0, \Delta t, 2\Delta t, \ldots,$ once the initial computations of $\exp \mathbf{A}\Delta t$, $\mathbf{B}$, $\mathbf{B}_{\mathbf{R}}$ and of the starting vector $\mathbf{Z}(0)$ have been made.

Finally we turn to determining the form of $X(t)$. Let the polynomial $P(z)$ be given by

$$P(z) = b_0 z^m + b_1 z^{m-1} + \ldots + b_m .$$

Then we see that

$$\begin{aligned} X(t) &= P(D)\, \phi(t) \\ &= b_0 Z_{m+1}(t) + b_1 Z_m(t) + \ldots + b_m Z_1(t), \end{aligned}$$

for $t = 0, \Delta t, 2\Delta t, \ldots,$ where $Z_i(t)$ is the ith component of $\mathbf{Z}(t)$.

The method outlined above has several pleasing features. First of all, it involves only knowledge of the power spectral density $S(\omega)$ and not the autocorrelation $R(\tau)$. This is extremely important in many engineering applications where one encounters $S(\omega)$ in a fairly natural way. Although the preliminary work involved in calculating the necessary matrices can be formidable, it need only be done once, and the amount of this work is quite independent of the sample size required (i.e., the number of sample points).

There are several places where additional information will streamline the computations. For example, if the matrix $\mathbf{A}$ is known to be diagonalizable, as is the case, for instance, when the roots of $Q(z)$ are all distinct, then the work involved in calculating $\exp \mathbf{A}\Delta t$ and $\mathbf{\Sigma}_{\mathbf{R}}$ can be drastically reduced. One simply uses the fact that for any non-singular matrix $\mathbf{T}$, $\exp(\mathbf{T}^{-1}\mathbf{A}\mathbf{T}\Delta t) = \mathbf{T}^{-1}(\exp \mathbf{A}\Delta t)\mathbf{T}$. If $\mathbf{T}^{-1}\mathbf{A}\mathbf{T}$ is diagonal then $\exp(\mathbf{T}^{-1}\mathbf{A}\mathbf{T}\Delta t)$ can be calculated simply by exponentiating the diagonal elements.

As a final remark, in generating the sequence $W_0, W_1, \ldots$ needed above, care must be taken to guard against unwanted correlation, since it is assumed that these numbers simulate numerically the white noise phenomena in the discrete case. In his paper [30] Franklin deals at length with the construction of a sequence with the desired properties.

Chapter 7

THE CALCULUS OF MONTE CARLO

1 Introduction to the Monte Carlo method

In its broadest interpretation the term "Monte Carlo" refers to a random simulation of a problem of a deterministic nature or to a direct simulation of a random process such as, for instance, a normal process as discussed in Chapter 6. In the case of a deterministic problem, such as the evaluation of an integral or the solution of a differential equation, one seeks a random process and a statistic whose expected value is numerically equal to the solution of the problem. For these methods to be valid, it is required that the statistic must converge to its expectation.

John von Neumann and Stanley Ulam seem to have been the first to have advocated the idea of systematically inverting the usual situation and treating deterministic problems by first finding a probabilistic analogue which is then solved by some sort of sampling procedure. In the early 1950's the Monte Carlo method was popularized and was soon being applied to an amazing variety of problems. Unfortunately, not enough thought was put into the design of the statistical model, and the method was applied in cases where failure was almost certain; the result of this was a general tendency to discredit the method. Although a better understanding of the capabilities of the method has since been obtained, there is nevertheless a lingering stigma attached to the term Monte Carlo. This is unfortunate since in many cases the Monte Carlo approach is at present the only feasible means of attacking a problem.

There are two main reasons for using Monte Carlo methods for solving deterministic problems. In the first place there are many problems of so complex a nature that no theoretical techniques are known for their solution. Secondly, at times a method of solution is known but is computationally unfeasible. An example of the latter which frequently occurs is the need to evaluate a multiple integral, say in as few as five dimensions. Except in the case of especially simple functions (integrands), existing numerical methods are very inefficient. A third, but less widely recognized, reason for resorting to Monte Carlo calculations is to obtain a rough estimate to use as a starting value in an iterative technique.

To give a rather naïve example of the use of Monte Carlo, let us examine various methods which could be used to evaluate

$$\theta = \int_0^1 x^2 \, dx.$$

On noticing that the graph of x^2 for $0 \leqslant x \leqslant 1$ is enclosed in the unit square, which has unit area, and using the interpretation of the integral as an area, we see that if $U = (U_1, U_2)$ is uniformly distributed on the square then θ is the probability that the point U lies below the graph of x^2. Thus, if out of N trials (that is, samples of U) we find that K points lie below the graph of x^2, then $\hat{\theta} = K/N$ is an estimate for θ.

Simple and intuitive as the above may seem, it is at the same time the most inefficient of the accepted Monte Carlo techniques. This method, including its obvious generalizations, is usually referred to as "hit-or-miss" Monte Carlo. It will be discussed later, but only for purposes of comparison with other methods. One can at least observe that rejection techniques for random-number generation essentially involve hit-or-miss Monte Carlo.

2 Methods of Monte Carlo integration

As the standard example we will consider the evaluation of the integral

$$\theta = \int_0^1 f(x) \, dx \tag{7.1}$$

in one dimension and will indicate later the obvious generalization to higher dimensions and other ranges of integration. From the statistical point of view, θ is the expected value of the random variable $f(U)$, where $U \sim U(0, 1)$. Thus it is immaterial whether we consider our problem from the point of view of integration or instead consider it as a statistical problem of estimating the parameter θ, at least as far as the numerical result is concerned.

However, if we take the second point of view, we may appeal to the following well-known result:

Theorem 7.1 If X is a random variable with distribution function $G(X)$ and expected value μ (finite) and finite variance, then the sample mean $\bar{X}$ of an independent sample $X_1, X_2, \ldots, X_n$ from G is an unbiased estimate of μ, that is $E(\bar{X}) = \mu$.

From Theorem 7.1 we can estimate θ in equation (7.1) by generating a sample $u_1, u_2, \ldots, u_n$ of uniform variates and setting

$$\hat{\theta} = \frac{1}{n} \sum_{i=1}^{n} f(u_i). \tag{7.2}$$

To obtain some idea of the accuracy of (7.2), we can compute the variance of $\hat{\theta}$, that is

$$\operatorname{var}(\hat{\theta}) = \frac{1}{n} \operatorname{var}(f). \tag{7.3}$$

Note that from equation (7.3) we see that the standard error in estimating θ is proportional to $1/\sqrt{n}$. As a matter of course, this holds true in any Monte Carlo calculation. The standard error is always of the form $e = k/\sqrt{n}$, where the constant k depends on the particular method used, and n is the number of random numbers used in the calculations. Moreover, the accuracy in doing m independent evaluations each using n numbers is the same as that obtained by doing a single computation involving mn numbers. Thus, to reduce the error by $\frac{1}{10}$, one must do 100 times as much work. Perhaps this explains why many of the early attempts at applying the Monte Carlo method ended in disaster. Clearly, if one is faced with the necessity of using Monte Carlo and having to accept the factor $1/\sqrt{n}$, it could be quite profitable to spend some time trying to devise a means of reducing the factor k. We will return to the problem of variance reduction later. Fortunately there are several techniques available which alleviate the slow convergence which is characteristic of Monte Carlo.

Before proceeding to a study of variance reduction, let us generalize our method to accommodate other integrands and ranges of integration. Let $f(p)$ be a real-valued function of an m-dimensional vector $p = (x_1, x_2, \ldots, x_m)$, and let $g(p)$ be a probability density function defined in R^m, the space of m-dimension real vectors. Denote by the less cumbersome notation

$$\theta = \int_{R^m} f(p)\, g(p)\, dv \tag{7.4}$$

the multiple integral

$$\int_{-\infty}^{\infty} \cdots \int_{-\infty}^{\infty} f(x_1, x_2, \ldots, x_m) g(x_1, \ldots, x_m) dx_1 \ldots dx_m.$$

Most of the integrals which one needs to evaluate can easily be put in the form (7.4), and sometimes in a quite natural and advantageous way. Now regardless of the original context, the integral in (7.4) can be interpreted as the expected value of a random variable $f(P)$, where P is a random vector with density function $g(p)$. Thus, Theorem (7.1) applies and we may estimate θ by

$$\theta = \frac{1}{n} \sum_{i=1}^{n} f(p_i) \tag{7.5}$$

where $p_1, p_2, \ldots, p_n$ are values of a random sample from the same distribution of P.

Exactly as before, we now have

$$\text{var}\,(\hat{\theta}) = \frac{1}{n}\,\text{var}\,(f) \tag{7.6}$$

where in this case the variance of f is

$$\text{var}\,(f) = \int_{R^m} (f(p) - \theta)^2 g(p) dv. \tag{7.7}$$

Since it would be quite unusual to be able to determine var (f) analytically, we estimate it from the points $p_1, p_2, \ldots, p_n$ used to determine $\hat{\theta}$ by the formula

$$s^2 = \frac{1}{n-1} \sum_{i=1}^{n} (f(p_i) - \hat{\theta})^2. \tag{7.8}$$

Thus, the same points used to estimate the integral give an estimate for the standard error, namely:

$$\hat{e} = s/\sqrt{n}. \tag{7.9}$$

A noteworthy feature of the Monte Carlo method of integration is the fact that the accuracy is relatively independent of the number of dimensions involved in the integral. While certain other numerical techniques for quadrature are more efficient for one-dimensional integrals, and even for some two- or three-dimensional integrals, they lose their appeal in higher dimensions, as the labour needed to achieve a given accuracy increases sharply. It is difficult to provide a decisive rule as to when to use Monte Carlo and when to use other methods, but at least a few guide-lines can be set forth. In the case of one-dimensional integrals, the Monte Carlo method is used only when the integrand is badly behaved and is discontinuous at several (unknown) points. For the two-dimensional case, the Monte Carlo method is to be preferred unless the integral can easily be written as a sum of integrals of differentiable functions. For higher dimensions one finds that Monte Carlo integration is comparable to the best numerical techniques available, except for the most trivial examples for which exact interpolation formulas are known.

3 Stratified sampling

In truth, the method to be presented here is a generalized version of that usually referred to as stratified sampling, which we obtain as a special case. For the remainder of this chapter we will deal only with one-dimensional integrals, since the notation is less cumbersome and the generalization to higher dimensions is fairly straightforward. Thus we suppose that (7.3) takes the form

$$\theta = \int_{-\infty}^{\infty} f(x) g(x) dx, \tag{7.10}$$

where $g(x)$ is a probability density function.

Now let us suppose that $g(x)$ can be written as a linear combination

$$g(x) = \sum_{i=1}^{k} p_i g_i(x), \tag{7.11}$$

where

$$0 \leqslant p_i \leqslant 1, \quad \sum_{i=1}^{k} p_i = 1,$$

and the $g_i(x)$ are also probability density functions. We recall, as in Chapter 3, that $g(x)$ is then the density function for a random variable X which is a mixture, with mixing probabilities p_i, of random variables $X_1, X_2, \ldots, X_k$, where each X_i is distributed according to $g_i(x)$. Let us set

$$\theta_i = \int_{-\infty}^{\infty} f(x) g_i(x) dx \tag{7.12}$$

and observe that from (7.10) and (7.11) we now have the identity

$$\theta = \sum_{i=1}^{k} p_i \theta_i . \tag{7.13}$$

As in the preceding section, we may consider each θ_i to be the expected value of the random variable $f(X_i)$. By applying the crude Monte Carlo method developed in that section we are able to find unbiased estimates $\hat{\theta}_i$ for the θ_i by generating n_i variates of X_i and using (7.5). Clearly,

$$\hat{\theta} = \sum_{i=1}^{k} p_i \hat{\theta}_i \tag{7.14}$$

is then an unbiased estimate of θ.

The utility of the above method is that by proper choice of the various quantities $g_i(x)$, p_i and the n_i, the variance of θ in (7.14) may be greatly reduced over that obtained by simply generating $n = n_1 + n_2 + \ldots + n_k$ variates of X and applying crude Monte Carlo, that is, (7.5), directly. To see more clearly how this is accomplished, let us set

$$\sigma^2 = \int_{-\infty}^{\infty} (f(x) - \theta)^2 g(x) dx \tag{7.15}$$

and

$$\sigma_i^2 = \int_{-\infty}^{\infty} (f(x) - \theta_i)^2 g_i(x) dx.$$

Hence, if n_i points are used to evaluate $\hat{\theta}_i$, then from (7.6) the variance is

$$\mathrm{var}\,(\hat{\theta}_i) = \sigma_i^2/n_i. \tag{7.16}$$

Now recall the following result of elementary statistics:

Theorem 7.2 If the random variable Y is given by

$$Y = \sum_{i=1}^{k} \alpha_i Y_i,$$

where $Y_1, Y_2, \ldots, Y_k$ are independent, then

$$E(Y) = \sum_{i=1}^{k} \alpha_i E(Y_i)$$

and

$$\mathrm{var}\,(Y) = \sum_{i=1}^{k} \alpha_i^2 \,\mathrm{var}\,(Y_i).$$

From the above theorem and the use of (7.14) and (7.16) it follows that

$$\mathrm{var}\,(\hat{\theta}) = \sum_{i=1}^{k} p_i^2\, \sigma_i^2/n_i\ . \tag{7.17}$$

It is this which is to be compared with the variance of the crude Monte Carlo estimate, which is σ^2/n for a comparable sample size. We are able to calculate σ^2 in terms of other quantities, and at times the expression is useful for comparing variances.

Theorem 7.3 With θ, θ_i, σ and σ_i as defined above, we have

$$\sigma^2 = \sum_{i=1}^{k} p_i\,[\sigma_i^2 + (\theta_i - \theta)^2]. \tag{7.18}$$

Proof Let us agree for simplicity that summations are over the range $1 \leqslant i \leqslant k$ and that integrations are over the range $-\infty < x < \infty$. Now we first observe that

$$\sum p_i\,(\theta_i - \theta)^2 = \sum p_i \theta_i^2 - \theta^2.$$

Now,
$$\begin{aligned}\sigma^2 &= \int (f(x) - \theta)^2 g(x)\,dx \\ &= \int [f(x)]^2 g(x)\,dx - \theta^2 \\ &= \Sigma p_i \int [f(x)]^2 g_i(x)dx - \theta^2 \\ &= \Sigma p_i (\sigma_i^2 + \theta_i^2) - \theta^2 \\ &= \Sigma p_i \sigma_i^2 + \Sigma p_i \theta_i^2 - \theta^2 \\ &= \Sigma p_i \sigma_i^2 + \Sigma p_i (\theta_i - \theta)^2 \\ &= \Sigma p_i [\sigma_i^2 + (\theta_i - \theta)^2],\end{aligned}$$
as desired.

In order to see what may be done to improve the possibility of decreasing the variance of $\hat{\theta}$, it is helpful to compare with crude Monte Carlo. Let θ' be the estimate of θ obtained by crude Monte Carlo with n points. Thus, var $(\theta') = \sigma^2/n$, from (7.3). Now using (7.18), after a little manipulation, we obtain

$$\text{var}(\theta') - \text{var}(\hat{\theta}) = \sum_{i=1}^{k} p_i (\theta_i - \theta)^2/n + \sum_{i=1}^{k} p_i \sigma_i^2 (1/n - p_i/n_i). \tag{7.19}$$

From this equation it is clear that a substantial reduction in variance can be obtained by choosing n_i approximately equal to np_i. In this case the second term in (7.19) vanishes and the variance is reduced by

$$\sum_{i=1}^{k} p_i (\theta_i - \theta)^2/n.$$

This, however, does not represent the best possible reduction of variance, as we see from our next theorem. Let us set $n_i = nq_i$, where $0 \leqslant q_i \leqslant 1$, and

$$\sum_{i=1}^{k} q_i = 1. \tag{7.20}$$

We may rewrite (7.20) as

$$\text{var}(\hat{\theta}) = \sum_{i=1}^{k} p_i^2 \sigma_i^2 / nq_i. \tag{7.21}$$

Theorem 7.4 The minimum value of var $(\hat{\theta})$ above is obtained when

$$q_i = p_i \sigma_i \bigg/ \left(\sum_{j=1}^{k} p_j \, \sigma_j \right). \tag{7.22}$$

Proof We use the method of Lagrangian multipliers in (7.21), subject to the constraint in (7.20). Thus we consider

$$\text{var}\,(\hat{\theta}) = \sum_{i=1}^{k} p_i^2 \sigma_i^2 / n q_i + \lambda \left(\sum_{i=1}^{k} q_i - 1 \right).$$

Then

$$\frac{\partial \text{ var}\,(\hat{\theta})}{\partial q_i} = -p_i^2 \sigma_i^2 / n q_i + \lambda$$

for each $i = 1, 2, \ldots, k$. Equating this to zero, we obtain

$$q_i = p_i \sigma_i / \sqrt{n\lambda}.$$

Now from the constraint in (7.20) we find that

$$\sqrt{n\lambda} = \sum_{i=1}^{k} p_i \sigma_i ,$$

which, together with the equation for q_i, gives (7.22). Clearly the critical point thus obtained is indeed a minimum.

Of course, in practice, Theorem 7.4 does not have much utility since the σ_i's are unknown and in all likelihood cannot be found analytically. However, they can be estimated during the course of a computation, and the number of sample points can be modified accordingly. The method presented above is very convenient when one is using a random number generator which actually mixes several distributions, such as the one given in Chapter 3 for the normal distribution. In this case the numbers $p_1, p_2, \ldots, p_k$ are just the mixing probabilities of the random number generator. This type of random number generator is easily devised, and therefore the above method is strongly recommended.

For Monte Carlo application, the distribution $g(x)$ is usually chosen as uniform on some interval $[a, b]$, and the decomposition into mixtures is usually no more than writing the integral (7.10) as a sum of integrals over disjoint intervals. That is, we are usually faced with an integral

$$\theta = \int_a^b f(x) \frac{1}{b-a} \, dx$$

where the $1/(b-a)$ factor has been introduced so that the above assumes the form of (7.12). We choose points $a = a_0 < a_1 < \ldots < a_k = b$ and write

$$\theta = \sum_{i=1}^{k} \frac{a_i - a_{i-1}}{b-a} \int_{a_{i-1}}^{a_i} f(x) \frac{1}{a_i - a_{i-1}} \, dx. \qquad (7.23)$$

Now this is in the form needed to apply the preceding methods with

$$P_i = \frac{a_i - a_{i-1}}{b-a}$$

and

$$g_i(x) = \begin{cases} \dfrac{1}{a_i - a_{i-1}} & \text{for } a_{i-1} \leqslant x \leqslant a_i \\ 0 & \text{otherwise.} \end{cases}$$

4 Control variates

One of the basic techniques for evaluating complicated integrals is to attempt to approximate the integrand by a simpler function which has a known integral. When this is used in conjunction with Monte Carlo methods, one can achieve greater success than would be possible by the use of either method alone. To see how this is done, suppose we wish to evaluate

$$\theta = \int_0^1 f(x)\,dx,$$

but the integrand cannot be integrated by theoretical means. Now if we can find a simpler function $h(x)$ which approximates $f(x)$ well on $[0, 1]$ and for which

$$\psi = \int_0^1 h(x)\,dx$$

is known, then we may write

$$\theta = \psi + \int_0^1 (f(x) - h(x))\,dx, \tag{7.24}$$

and if the approximation is good, the second term above will be small. Actually, if we intend to evaluate the integral in (7.24) by Monte Carlo methods, we need var $(f(x) - h(x)) <$ var $(f(x))$. Clearly,

$$\text{var}\,(f(x) - h(x)) = \text{var}\,(f(x)) + \text{var}\,(h(x)) - 2\,\text{cov}\,(f(x), h(x)),$$

where

$$\text{cov}\,(f(x), h(x)) = \int_0^1 (f(x) - \theta)\,(h(x) - \psi)\,dx.$$

Hence we need

$$\text{var}\,(h(x)) < 2\,\text{cov}\,(f(x), h(x)). \tag{7.25}$$

We obtain estimates $\hat{\theta}$ and $\hat{\psi}$ for θ and ψ by using (7.2). The strong positive correlation is introduced by using the same random numbers to evaluate $\hat{\theta}$ and $\hat{\psi}$. Assuming that (7.23) is satisfied, we have

$$\begin{aligned} \text{var}\,(\psi + \hat{\theta} - \hat{\psi}) &= \text{var}\,(\hat{\theta} - \hat{\psi}) \\ &= \text{var}\,(f(x) - h(x))/n < \text{var}\,(f(x))/n \\ &= \text{var}\,(\hat{\theta}). \end{aligned} \tag{7.26}$$

Although as presented above the method is primarily for integration, where $h(x)$ is called the "control variate," it may be generalized to other Monte Carlo applications. In many instances a single Monte Carlo run provides as output an estimator $\hat{\theta}$ whose expected value is equal to that of an unknown parameter θ. We seek another estimator $\hat{\psi}$, here again called the control variate, whose expected value is a known constant ψ, and which has a strong positive correlation with $\hat{\theta}$. The quantity

$$\theta' = \hat{\theta} + \psi - \hat{\psi} \tag{7.27}$$

is also an unbiased estimator of θ. We have

$$\text{var}\,(\theta') = \text{var}\,(\hat{\theta}) + \text{var}\,(\hat{\psi}) - 2\,\text{cov}\,(\hat{\theta}, \hat{\psi})$$

and if var $\hat{\psi} < 2 \operatorname{cov}(\hat{\theta}, \hat{\psi})$, then

$$\operatorname{var}(\theta') < \operatorname{var}(\hat{\theta}). \tag{7.28}$$

This is perhaps the strongest of all techniques for reducing the variance in Monte Carlo calculations, especially in cases where the desired positive correlation is easily introduced, as for example in integration.

5 Antithetic variates

Again, suppose that $\hat{\theta}$ is an unbiased estimator of some parameter θ. Assume further that we have a second unbiased estimator θ' of θ, which has a strong negative correlation with $\hat{\theta}$. The quantity

$$\hat{\psi} = (\hat{\theta} + \theta')/2 \tag{7.29}$$

is also an unbiased estimator of θ, and

$$\operatorname{var} \hat{\psi} = \tfrac{1}{4} \operatorname{var}(\hat{\theta}) + \tfrac{1}{4} \operatorname{var}(\theta') + \tfrac{1}{2} \operatorname{cov}(\hat{\theta}, \theta'). \tag{7.30}$$

It is easily seen from (7.30) that var $(\hat{\psi})$ can at times be made smaller than var $(\hat{\theta})$ by a suitable choice of θ'.

To illustrate the typical situation where the method can be applied, suppose that $f(x)$ is an increasing function which is to be integrated from 0 to 1. If U is uniformly distributed on $[0, 1]$, then U and $1 - U$ have a correlation of -1. It is easily seen that since f is increasing, $f(U)$ and $f(1 - U)$ are negatively correlated, that is, antithetic. Moreover, they each have the same expected value

$$\theta = \int_0^1 f(x)\,dx,$$

the desired quantity. In this case we can take $\hat{\theta}$ to be the crude Monte Carlo estimate of θ defined as

$$\hat{\theta} = \left(\sum_{i=1}^{n} f(u_i) \right) \Big/ n,$$

and take θ' to be

$$\theta' = \left(\sum_{i=1}^{n} f(1 - u_i) \right) \Big/ n,$$

where $u_1, u_2, \ldots, u_n$ are generated from $U(0,1)$. Observe that the negative correlation is introduced by using the same random numbers in each case.

The antithetic variate method applied to such problems as above is intuitively appealing when interpreted in the following sense: $\hat{\psi}$ is the crude Monte Carlo estimate of the integral

$$\theta = \int_0^1 g(x)\,dx,$$

where $g(x) = (f(x) + f(1-x))/2$. The assumptions on $f(x)$ assure us that the symmetrized function $g(x)$ will have a smaller variance than that of $f(x)$.

In situations where it is easily and naturally applied, the antithetic variate method is very efficient. However, it is not easily generalized to higher dimensions when the purpose is the evaluation of multiple integrals, and it is therefore not as useful for integration as the other techniques presented so far.

6 Remarks

A good basic premise for Monte Carlo integration – in fact for any Monte Carlo calculation – is never to solve a problem in its natural form just for the sake of being direct. In many instances it pays to search for a different problem with the same numerical answer but which is more amenable to solution. If at any point in a calculation a particular portion can be done analytically, then to do so will ultimately lead to a more accurate result, that is, a reduction of variance.

If a certain Monte Carlo procedure requires an amount of work w (measured in some convenient system of units, such as computing time) and the resulting estimate has a variance of σ^2, then we define the deficiency by

$$d = w\sigma^2. \tag{7.31}$$

Evidently, the deficiencies of two procedures can be compared by examining the ratio

$$d_1/d_2 = (w_1\sigma_1^2)/w_2\sigma_2^2 = \left(\frac{w_1}{w_2}\right)\left(\frac{\sigma_1^2}{\sigma_2^2}\right) \tag{7.32}$$

even though the deficiency d is relatively meaningless. Note that (7.32) is the product of two terms, namely w_1/w_2, called the "labour ratio," and σ_1^2/σ_2^2, called the "variance ratio". The labour ratio is usually dependent on available computing apparatus, so that estimates of this ratio will depend on factors other than the immediate problem. On the other hand, the variance ratio depends only on the particular methods and can often be provided theoretically. For example, consider the evaluation of the integral of an increasing function $f(x)$ by means of antithetic variates, where the variance, say σ_1^2, is given by (7.30). Comparing this with crude Monte Carlo, with variance σ_2^2, we have

$$\sigma_1^2/\sigma_2^2 = \tfrac{1}{2} + \tfrac{1}{2}\rho, \qquad (7.33)$$

where ρ is the correlation coefficient between $f(U)$ and $f(1-U)$ for $U \sim U(0,1)$. Now a generous estimate for the labour ratio is $w_1/w_2 = 2$. In fact, the antithetic variate method requires twice as many evaluations of $f(x)$ but the same number of random numbers, so the work is slightly less than twice as much. In any case, this gives a relative deficiency

$$d_1/d_2 = 1 + \rho.$$

We recall that the purpose was to introduce a negative correlation, and if this is accomplished, we will have $d_1/d_2 < 1$, as desired. By the same token, if we have failed to achieve a negative correlation, then we have $d_1/d_2 > 1$, so that crude Monte Carlo is the better method in this case.

For those who are familiar with the theory of sample surveys, it should be noted that the development of stratified sampling in section 3, above, is not unlike the developments of proportional allocation and of optimal allocation in sampling. The similarities are striking, especially when the specific development here is compared with the general development of sampling theory in [65, Vol. 3, pp. 179–182].

Also, one should note the similarities of the development of the control variable to that of ratio estimation in sample survey theory [65, Vol. 3, p. 216] and the method of using supplementary information as reported in [26].

However, for those not familiar with sample survey theory, that which has been developed here is sufficient for understanding the principles as far as concerns Monte Carlo methods.

Chapter 8

SOLUTION OF LINEAR PROBLEMS

1 Introduction

In the area of linear problems, existing Monte Carlo techniques are for the most part very inefficient. There seem to be only a few situations in which there is anything to be gained by the use of Monte Carlo over conventional methods. Conventional methods are to be preferred except when a rough estimate of the solution is required as a starting point for later work, or when the problem is too large or intricate to be treated in any other fashion. Nevertheless, there is something to be learned from considering the techniques used to solve linear equations. In accordance with the above remarks, the presentation which follows will be rather superficial, since it is the general idea of the methods and not the specific details which find important applications in other types of problems.

We shall be concerned in this chapter with two basic problems. The first of these is the problem of solving a system of simultaneous linear equations which we can write in matrix notation as

$$\mathbf{A}\mathbf{x} = \mathbf{b}$$

where $\mathbf{A}$ is an $n \times n$ matrix and $\mathbf{b}$ is a column vector. A more convenient form of the above is obtained by substituting $\boldsymbol{H} = \boldsymbol{I} - \mathbf{A}$ to obtain the iterative form

$$\mathbf{x} = \mathbf{b} + \boldsymbol{H}\mathbf{x}.$$

Along this same line, the latter form suggests the problem of finding a solution $f(x)$ of the integral equation

$$f(x) = g(x) + \int K(x, y) f(y) dy ,$$

which is a Fredholm integral equation of the second kind.

The second problem, which is one of the most popular illustrations of Monte Carlo, is the solution of the Dirichlet problem in potential theory. Given a closed domain D with boundary C and a prescribed function $f(x)$ defined on C, the problem is to find a function u which is continuous and differentiable on D satisfying

$$\nabla^2 u = 0$$

and

$$f(x) = u(x) \quad \text{for } x \in C.$$

It is perhaps unfortunate that this problem is so often put forward to illustrate the Monte Carlo approach, since the method of solution, while easy to grasp, is very inefficient as compared with other methods such as relaxation methods.

2 Linear equations

Let us consider a system of n linear equations in n unknowns, which we write

$$\mathbf{A}\mathbf{x} = \mathbf{b}. \tag{8.1}$$

As before, we substitute $\boldsymbol{H} = \boldsymbol{I} - \boldsymbol{A}$ and write the equations in the form

$$\mathbf{x} = \mathbf{b} + \boldsymbol{H}\mathbf{x}. \tag{8.2}$$

The Monte Carlo method of solving (8.2) is essentially due to von Neumann and Ulam, as reported by Taussky and Todd [99]. The elements h_{ij} of $\boldsymbol{H}$ are first factored in the form

$$h_{ij} = v_{ij}\, p_{ij} \tag{8.3}$$

where

$$p_{ij} \geqslant 0, \qquad \sum_{j=1}^{n} p_{ij} < 1. \tag{8.4}$$

The matrix $\boldsymbol{P} = (p_{ij})$ describes a terminating random walk on the states $1, 2, \ldots, n$ with transition probabilities p_{ij} and stop probabilities

$$p_i = 1 - \sum_{j=1}^{n} p_{ij}. \tag{8.5}$$

Let $\boldsymbol{\alpha} = (i_0, i_1, \ldots, i_k)$ denote a walk starting at i_0 and terminating

k steps later in the state i_k. Define

$$V_m(\boldsymbol{\alpha}) = v_{i_0 i_1} v_{i_1 i_2} \cdots v_{i_{m-1} i_m}, \quad m \leq k. \tag{8.6}$$

and also,

$$X(\boldsymbol{\alpha}) = (V_k(\boldsymbol{\alpha}) b_{i_k}) / p_{i_k} \tag{8.7}$$

where b_r is the rth component of $\mathbf{b}$. The probability that the walk $\boldsymbol{\alpha}$ occurs, given the starting value $i_0 = i$, is

$$\Pr(\boldsymbol{\alpha}) = p_{ii_1} p_{i_1 i_2} \cdots p_{i_{k-1} i_k} p_i .$$

Restricting to the walks beginning at $i_0 = i$, we see that the expectation of $X(\boldsymbol{\alpha})$ is given by

$$\sum_{\boldsymbol{\alpha}} \Pr(\boldsymbol{\alpha}) X(\boldsymbol{\alpha}) =$$

$$\sum_{k=0}^{\infty} \sum_{i_1} \cdots \sum_{i_k} (p_{ii_1} p_{i_1 i_2} \cdots p_{i_{k-1} i_k} p_{i_k} v_{ii_1} \cdots v_{i_{k-1} i_k} b_{i_k} / p_{i_k})$$

$$= \sum_{k=0}^{\infty} \sum_{i_1} \cdots \sum_{j_k} h_{ii_1} h_{i_1 i_2} \cdots h_{i_{k-1} i_k} b_{i_k}$$

$$= b_i + (\boldsymbol{H}\mathbf{b})_i + (\boldsymbol{H}^2\mathbf{b})_i + \ldots, \tag{8.8}$$

where the innermost sums are from 1 to n, and $(\boldsymbol{H}^m\mathbf{b})_i$ is the ith component of $\boldsymbol{H}^m\mathbf{b}$. The question of convergence of (8.8) is tantamount to the convergence of the Neumann series [34]

$$\boldsymbol{I} + \boldsymbol{H} + \boldsymbol{H}^2 + \ldots, \tag{8.9}$$

which converges if and only if the characteristic values of $\boldsymbol{H}$ lie inside the unit sphere in the complex plane; and in this case (8.9) converges to $\mathbf{A}^{-1}$. If the series in (8.9) does converge, then we have

$$x_i = (\mathbf{A}^{-1}\mathbf{b})_i = \mathrm{E}(X(\boldsymbol{\alpha})), \tag{8.10}$$

conditional on having $i = i_0$. We may therefore estimate any component x_i by simply generating random walks $\boldsymbol{\alpha}$ starting from $i_0 = i$ and scoring $X(\boldsymbol{\alpha})$ when the walk terminates. The estimate for x_i is then obtained by averaging the scores of the walks.

In 1956 Curtiss [19] compared the efficiency of the Monte

Carlo method with conventional methods for computing a component of the solution vector **x**. Defining the norm of the matrix H by

$$||H|| = \max_i \left(\sum_j |h_{ij}| \right) \tag{8.11}$$

he found that

(i) if $||H|| > 1$, the Monte Carlo method fails;

(ii) if $||H|| = 0{\cdot}9$, then Monte Carlo is less efficient than a conventional method in obtaining 1 per cent accuracy with $n \leqslant 554$ or 10 per cent accuracy with $n \leqslant 84$;

(iii) if $||H|| = 0{\cdot}5$, then these figures are $n \leqslant 151$ for 1 per cent accuracy and $n \leqslant 20$ for 10 per cent accuracy.

From the above points it is apparent that the Monte Carlo method should not be used for most problems. On the other hand there are a few desirable features which should be stressed. In the first place, by taking **b** in (8.1) to be the jth unit vector the method, applied to the starting value $i_0 = i$, provides the (i, j)th element of $\mathbf{A}^{-1}$. Thus, Monte Carlo provides a way of estimating a single element or any collection of the elements of $\mathbf{A}^{-1}$. This is a very important feature since, for example, one often needs to know the diagonal elements of $\mathbf{A}^{-1}$ but does not wish to go to the bother of computing all of the inverse. Secondly, the amount of work involved in a computation is almost proportional to n, and not to n^3 as is the case with most methods of solving n equations in n unknowns.

Very little is needed to adapt the preceding method to handle Fredholm integral equations of the second kind, such as

$$f(x) = g(x) + \int K(x, y) f(y) dy. \tag{8.12}$$

We simply replace the integral above by a quadrature formula on n points and solve the resulting matrix equation at the quadrature points. The Monte Carlo method could be very advantageous here if a prohibitive number of quadrature points need to be used. Note that the method can be used to yield the value of f at a single point instead of having to calculate the value at all of the quadrature points.

3 The Dirichlet problem

To illustrate the application of Monte Carlo to boundary value problems, we shall consider the Dirichlet problem for Laplace's equation in two dimensions. Thus we consider a function f defined on the boundary C of a closed domain D in two dimensions. We wish to find a solution of

$$\nabla^2 u(x, y) = \frac{\partial^2}{\partial x^2} u(x, y) + \frac{\partial^2}{\partial y^2} u(x, y) = 0, \tag{8.13}$$

subject to the conditions

$$u(x, y) = f(x, y)$$

on the boundary C. We begin by covering D with a grid having equal spacing between lines equal to h. The points of the grid are the intersections of the horizontal and vertical lines and can be divided into two types, namely those with four neighbouring points interior to D and called "interior" points, and those with fewer than four neighbouring points interior to D, called "boundary" points.

We assume that for each boundary point Q the function u assumes the value $f(Q)$. We attempt to find the values of u at interior points by replacing (8.13) by its finite difference analogue. If P_1, P_2, P_3, P_4 denote the four adjacent points to the interior point P, then the finite difference Laplacian will be

$$h^{-2}[u(P_1) + u(P_2) + u(P_3) + u(P_4) - 4u(P)]. \tag{8.14}$$

Equating (8.14) to zero, we obtain

$$u(P) = \tfrac{1}{4}[u(P_1) + \dots + u(P_4)] \tag{8.15}$$

at the interior points P, and of course

$$u(Q) = f(Q) \tag{8.16}$$

at the boundary points Q. The essence of the method is to solve the system of linear equations defined by (8.15) and (8.16). As a matter of interest, it is easy to see that these equations are given in the form of (8.2). The elements of $\boldsymbol{H}$ are all either 0 or $\frac{1}{4}$, and in the case where there is at least one interior grid point then from (8.11) we see that $||\boldsymbol{H}|| = 1$. In view of (i) following (8.11) it is not surprising that the Monte Carlo method is inefficient here. In fact, the Neumann series for $\boldsymbol{H}$ must diverge. Actually, however, the method of solution is simpler and more direct than those in section 2.

Let us do a terminating random walk beginning at a point P and ending immediately upon reaching a boundary point Q (ending at once if P is itself a boundary point). We assume that for each interior point P the probability of transition to any adjacent point is the same and equal to $\frac{1}{4}$. Let $u(P, Q)$ denote the probability that a walk beginning at a point P terminates at the boundary point Q. From elementary probability we have at once

$$u(P, Q) = \tfrac{1}{4} \sum_{i=1}^{4} u(P_i, Q). \tag{8.17}$$

Moreover, $u(P,Q)$ satisfies the conditions

$$u(Q, Q) = 1 \tag{8.18a}$$

and

$$u(Q', Q) = 0 \tag{8.18b}$$

where $Q' \neq Q$ is also a boundary point. Note that, as a function of P, $u(P, Q)$ satisfies (8.15) but the boundary conditions are wrong. We now proceed to effect a conversion of the boundary conditions.

Let us suppose that upon terminating a walk at Q a score of $f(Q)$ is tallied. The score $\tau(P)$ tallied after starting a walk at the point P is a random quantity assuming the values $f(Q_1)$, $f(Q_2), \ldots, f(Q_s)$, where $Q_1, Q_2, \ldots, Q_s$ exhaust the boundary points. The probability that $\tau(P) = f(Q_j)$ is obviously $u(P, Q_j)$, so that the expected value of $\tau(P)$ is given by

$$w(P) = \mathrm{E}(\tau(P)) = \sum_{j=1}^{s} f(Q_j) u(P, Q_j). \tag{8.19}$$

Now we see that

$$\begin{aligned} \tfrac{1}{4} \sum_{i=1}^{4} w(P_i) &= \tfrac{1}{4} \sum_{i=1}^{4} \sum_{j=1}^{s} f(Q_j) u(P_i, Q_j) \\ &= \sum_{j=1}^{s} f(Q_j) \left(\frac{1}{4} \sum_{i=1}^{4} u(P_i, Q_j) \right) \\ &= \sum_{j=1}^{s} f(Q_j) u(P, Q_j) = w(P), \end{aligned}$$

making use of (8.17). That is, $w(P)$ satisfies (8.15). Moreover, using (8.18a and b) it follows that

$$w(Q) = \sum_{j=1}^{s} f(Q_j)u(Q, Q_j) = f(Q),$$

so that $w(P)$ also satisfies (8.16). Thus $w(P)$ is a solution of the Dirichlet problem or, more precisely, of the finite difference approximation.

It only remains to find estimates for $u(P, Q_j)$, $j = 1,2,\ldots, s$. Suppose that out of N walks initiated at P, N_j of these terminate at Q_j. Clearly the ratio N_j/N is an unbiased estimate of $u(P, Q_j)$. Hence (replacing $w(P)$ by $u(P)$),

$$u(P) = \frac{1}{N} \sum_{j=1}^{s} N_j f(Q_j) \tag{8.20}$$

is an unbiased estimate for the solution of (8.15) and (8.16).

To get some idea of the efficiency of the above method, it is helpful to estimate the time T required for obtaining the solution to within an error ϵ. If the time required for a single step in a walk is t and if ν_i is the number of steps in the ith walk, then

$$T = t\left(\sum_{i=1}^{N} \nu_i\right).$$

Since the number of steps ν in a walk is a random quantity and since N will probably be large, we are justified in writing

$$T = tN\,\mathrm{E}(\nu)$$

where $\mathrm{E}(\nu)$ is the expected value of ν. Now N may be determined from the law of large numbers, and we obtain

$$N = 9\sigma^2/\epsilon^2,$$

where σ^2 is the variance of $\tau(P)$. σ^2 may be estimated as the computations progress, or we may certainly use the upper bound

$$\sigma^2 \leqslant \max_j |f(Q_j)|^2 - \min_j |f(Q_j)|^2.$$

The quantity $\mathrm{E}(\nu)$, as it turns out, depends only on the linear dimensions of the grid [9]. For instance, if D is a spherical region of radius r (where units are chosen to make $h = 1$) then $E(v)$ is proportional to r^2. It is an important fact that this is independent of the number of dimensions in the problem. In general, the time needed for a solution is proportional to tr^2/ϵ^2. The accuracy, in addition, will depend on the coarseness of the grid. As an example, if the solution is desired to an accuracy of $\epsilon = \cdot 01$ of the maximum value of $f(Q)$, then $r = 100$, $N = 10^4$ and $T = 10^8 t$. For $t = 100$ microseconds, this gives T to be about three hours.

There are various refinements which make the above method more efficient. Suppose a walk beginning at a point P revisits the point $k - 1$ times before finally reaching the boundary at a point Q. We may consider this as k walks beginning at P and ending at Q. Thus with N and N_j as in (8.20), if $Q = Q_j$ we increase N and N_j by k.

Although the preceding was illustrated for the two-dimensional problem, the generalization to higher dimensions is obvious. One should again note that the Monte Carlo method does not get increasingly more difficult as the number of dimensions is increase Also, as with the linear equation problem, we are again able to calculate the solution at a single point and do not have to evaluate the solution at every mesh point.

Chapter 9

TESTING RANDOM NUMBER GENERATORS

In this chapter we intend to present some of the tests of random numbers which are traditional with statisticians, and to make some general recommendations about other tests. It is a fact that specific tests must be designed to fit a particular problem if we are to have any assurance that the results will be meaningful. Nevertheless, it is possible to eliminate bad choices of random number generators on the basis of failure to have certain desirable properties common to any source of random numbers. We shall confine our attention to generators providing samples from a uniform distribution on [0, 1], since in practice other distributions are obtained from this by some kind of conversion process. In addition, we shall be primarily concerned with the aspects of computation on a modern binary computer. Thus, special attention will be paid to the fact that computations will ordinarily be carried out on high-speed binary equipment. The tests discussed in this chapter are statistical, that is, by their nature a null hypothesis is necessarily involved. This in turn means that one must select an appropriate level of significance. Also, if one wishes to discuss the power of such tests, it is necessary to define a meaningful alternative hypothesis. For example, when testing to determine whether a sample could have come from a uniform density with known parameters, the null hypothesis is automatically defined; however, the question whether this hypothesis is being tested versus a known alternative – such as that the alternative probability density is normal, chi-square, or uniform with different and perhaps unknown parameters – is difficult to resolve. What we are trying to say is that, realistically, the most one can do is simply to state that the sample is either likely or unlikely to have come from the desired uniform probability density.

The choice of significance level for most tests appearing in the literature is an arbitrary choice dictated more by custom than by an objective mathematical analysis. Since sample sizes are generally large, one may choose the significance level small, say 0·01, without great concern as to whether the effect on the power is detrimental, except perhaps in some pathological cases.

A system of tests was devised by Kendall and Babington Smith [62] to check for what they called "randomness". The system consisted of four tests : the frequency test, the serial test [62], [35], [36], the poker test, and the gap test. A set of random numbers "passing" all these tests is called "random". Several of these tests will be discussed later in this chapter and in more detail, but the rationale is given here for motivation. The first test, the frequency test, is applied to ensure that all digits shall occur approximately an equal number of times. The second test, the serial test, is applied to ensure that no two digits shall tend to re-occur in sequence; therefore, if one formed a bivariate table showing the distribution of pairs of digits in the series, arranged in rows according to the first digit and in columns according to the second digit, one should obtain frequencies which are approximately equal in all the cells. In the poker test the digits are arranged in blocks, say of five each, so as to form a "poker" hand, and the observed frequencies of occurrence of the different hands, such as four of a kind, two pairs, etc., are compared with the theoretical values. Finally, the gap test measures the expectation of the length of the gap between successive occurrences of a given digit and compares this with the theoretical value. In practice the gap test is usually replaced by a frequency test of the length of runs [94].

One notes immediately that the re-use of the same data in testing implies that these tests are correlated in some manner. Even so, Kendall and Babington Smith believe that a set of numbers that passes these four tests will have the desirable properties that random numbers heuristically should have. Apparently, what has actually been formulated, then, is a definition of a set of generated random numbers as a set of numbers that can "pass" a sequence of statistical tests judiciously selected by the experimenter. This idea is compatible with D.H. Lehmer's definition [66] of a sequence of random numbers :

"...[a] vague notion embodying the idea of a sequence in which each term is unpredictable to the uninitiated and whose digits pass a certain number of tests traditional with statisticians and depending somewhat on the use to which the sequence is to be put."

The chi-square statistic for goodness-of-fit, the Kolmogorov–Smirnov and the W^2 statistics are used to test whether a sequence of random numbers is from a uniform distribution with known parameters. Each of these tests is discussed briefly in this chapter.

1 The frequency test

The frequency test should be applied to every segment of a sequence to be used in a given computation. This is feasible by its very simplicity. The reason for this is that even with a random number generator which is known to be uniform over a full period, such as one of the congruential generators, a particular segment may be very bad.

The test is very simple. One simply divides the interval [0, 1] into k non-overlapping subintervals and counts the number n_i of times that a number from a sequence of length n falls in the ith interval. These are then compared with the expected values by some standard test such as the χ^2 test. The expected number e_i is just n times the length of the ith interval. One then computes the quantity

$$\chi^2 = \sum_{i=1}^{k} \frac{(n_i - e_i)^2}{e_i} \tag{9.1}$$

and consults a table for the χ^2 distribution with $k - 1$ degrees of freedom. Since too close an agreement can be just as suspect as too much deviation, it is typical to reject if the probability of more deviation is greater than ·95 or less than ·05, the latter in any case. In practice, one usually takes equally spaced subdivisions of length $1/k$, so that $e_i = n/k$. In this case we have

$$\chi^2 = k\left(\sum_{i=1}^{k} n_i^2\right) - n. \tag{9.2}$$

For binary computers it is most convenient to take k to be a power of 2. The proper interval can be obtained simply by a right shift in this case, assuming that the numbers are in unnormalized integer form.

Note that we may interpret the frequency test as a way of

determining the usability of the generator as a source of random values of the integers $0, 1, 2, \ldots, k$; something which is essential for problems involving random walks.

2 Tests for goodness of fit

One of the objections raised at times to the χ^2 test given above is that it is relatively insensitive. In this section we shall discuss two tests for goodness of fit to a particular distribution. Both of these are somewhat difficult to apply correctly since they require that the sample be ordered in, say, ascending order.

The first of these is the Kolmogorov–Smirnov test [65, Vol. 2]. Suppose that $x_1, x_2, \ldots, x_n$ is a sample of size n, which has been ordered $x_i < x_{i+1}$, coming from a continuous distribution with distribution function $F(x)$. Of course, we are assuming that the distribution is $U(0, 1)$, but the test is distribution-free. We define the sample distribution function $S_n(x)$ by

$$S_n(x) = k/n \quad \text{where } x_k \leqslant x < x_{k+1}. \tag{9.3}$$

The test is based on the statistic

$$D_n = \sup_n |S_n(x) - F(x)|. \tag{9.4}$$

The asymptotic distribution for D_n was obtained by Kolmogorov in 1933 and has been extensively tabulated. Due to the nature of $S_n(x)$ and $F(x)$, the supremum in (9.4) will occur at one of the sample points, so that D_n is the maximum of all expressions of the form

$$|S_n(x_i) - F(x_i)|$$

or

$$|S_n(x_i) - F(x_{i+1})|$$

(for $i = n$ we take $x_{n+1} = +\infty$).

To apply the test, the quantity $\sqrt{n}\, D_n$ is compared with values from a table and the hypothesis is rejected, that is the sample is rejected, if the value of $\sqrt{n}\, D_n$ is in excess of certain critical values. For convenience we list the values for the ·01 and ·05 level of significance.

Level of significance	Critical value of $\sqrt{n}\, D_n$
·05	1·3581
·01	1·6276

These values are adequate for use when $n \geqslant 80$.

The second of these tests is the W^2 test, based on the quantity

$$W^2 = \mathrm{E}((S_n(X) - F(X))^2)$$

$$= \int_{-\infty}^{\infty} (S_n(x) - F(x))^2 dF(x). \qquad (9.5)$$

A development of the theory associated with W^2 is given in [65]. W^2 is also distribution-free. Still assuming the sample to be ordered, we may write

$$nW^2 = \frac{1}{12n} + \sum_{k=1}^{n} \left(F(x_k) - \frac{2k-1}{2n} \right)^2 \qquad (9.6)$$

For the special case in which the distribution is $U(0, 1)$ so that $F(x) = x$ for $0 \leqslant x \leqslant 1$, we have

$$nW^2 = \frac{1}{12n} + \sum_{k=1}^{n} \left(x_k - \frac{2k-1}{2n} \right)^2 \qquad (9.7)$$

For test purposes the quantity nW^2 is compared with tabular values, and the sample is rejected if it exceeds the tabulated value. A list of the more commonly used values is given below:

Level of significance	Critical values of nW^2
·10	·347
·05	·461
·01	·743
·001	1·168

These values are actually for the asymptotic distribution, but it is shown in [72] that they are adequate for n as small as 3.

Although the above tests are time-consuming, they are very sensitive, and thus are recommended for all critical work. It is possible to obtain lower bounds for these two statistics from the frequency table used in section 1 in a simple and straightforward manner. At times, these lower bounds can be a motive for rejecting a sample.

3 The run test

Consider a sequence $x_0, x_1, x_2, \ldots, x_n$. In the case where

$x_p > x_{p+1} < x_{p+2} < \ldots < x_{p+k} > x_{p+k+1}$, we call x_{p+1}, $x_{p+2}, \ldots, x_{p+k}$ an upward run of length k. A downward run is defined by reversing all equalities. Thus, in the sequence

$$\overline{(1),\ \underline{5,}}\ \underline{3,}\ \overline{\underline{2},\ 3,\ 4,\ 5,\ 6}\ ,$$

where runs up are overscored and runs down are underscored, we have runs of length 1, 2, and 4. Note that the starting value x_0 indicated in parentheses above, where $x_0 = 1$, is not considered part of the sequence. The expected number of runs of length k in a sequence of length n can be shown [57] to be given by

$$R(k, n) = \frac{2[(k^2 + 3k + 1)n - (k^3 + 3k^2 - k - 4)]}{(k + 3)!} \qquad (9.8)$$

The run test consists of computing a value of χ^2 for runs of length less than or equal to k, and comparing with a table for the χ^2 distribution with $k - 1$ degrees of freedom. It is indicated in [33] that this test has good discrimination. Since it is easy to apply, it is reasonable to include it in any test package.

4 The serial test

Up to this point the only test we have considered which is concerned with order in the sequence has been the run test. Another such test is the serial test, which is in a sense a test of independence between x_{j-1} and x_j. Let us divide the interval [0, 1] into k equally spaced subintervals and count the number of times n_{ij} that a value x_{p-1} in the ith interval is followed by x_p in the jth interval. This gives a $k \times k$ matrix $\mathbf{N} = (n_{ij})$, and under the assumptions of independence and uniformity the elements of $\mathbf{N}$ all have the expected value n/k^2. Good shows in [35] that if

$$n_i = \sum_{j=1}^{k} n_{ij} ,$$

then

$$(k^2/n) \sum_{i=1}^{k} \sum_{j=1}^{k} (n_{ij} - n/k^2)^2 - (k/n) \sum_{i=1}^{k} (n_i - n/k)^2 \qquad (9.9)$$

is approximately distributed as χ^2 with $k(k - 1)$ degrees of freedom. This test is known as the serial test.

5 Higher dimensional tests

In the case where the sequence $x_0, x_1, \ldots, x_{nk}$ is intended to produce a sample of n vectors in k dimensions by means of defining

$$y_{p+1} = (x_{pk+1}, x_{pk+2}, \ldots, x_{pk+k}) \tag{9.10}$$

one should provide some means of testing for equidistribution on the k-cube. A subdivision of $[0, 1]$ into r equally spaced intervals defines a corresponding subdivision of the k-cube into r^k equally spaced cells. Clearly, for very large k this becomes impractical, since for 10 per cent accuracy $r = 10$, and for 1 per cent accuracy $r = 100$, the latter being impractical even for $k = 2$.

There are two alternatives, which are not entirely satisfactory but will at least give some idea of the quality of the random vectors produced. In the first case, one can examine the lengths of runs in the individual vectors and devise a test based on these. For instance, the average value of the number of runs of length r in a k-vector is an estimate of $R(r, k)$ from equation (9.8). Thus the accuracy of the ability to estimate $R(r, k)$ by Monte Carlo is a rough measure of the merits of the generator. For $k \leqslant 5$ there is the possibility of counting the number of vectors in each order type and comparing with the theoretical value of $n/k!$ that is, there are $k!$ distinct ways in which k numbers can be ordered. Under the assumption of independence these are equally likely (of course, the converse is not true). The observed values can be compared with the expected values by means of a χ^2 table with $k! - 1$ degrees of freedom.

Apparently, however, the testing of generators of random vectors would best be carried out by constructing a simulation problem to which the solution is known, and which requires essentially the same characteristics as the problem which is ultimately to be solved.

BIBLIOGRAPHY

[1] Anderson, T.W., *An Introduction to Multivariate Statistical Analysis*, John Wiley and Sons, 1958.

[2] Barnett, V.D., The behavior of pseudo-random sequences generated on computers by the multiplicative congruential method, *Math. Comp.*, **16** (1962), 63–9.

[3] Bispham, J.W., An experimental determination of the distribution of the partial correlation coefficient in samples of thirty, *Proc. Roy. Soc. London*, **97** (1920), 218–24.

[4] Bispham, J.W., An experimental determination of the distribution of the partial correlation coefficient in samples of thirty, *Metron*, **2** (1923), 684–96.

[5] Blackman, R.B. and Tukey J.W., *The Measurement of Power Spectra from the Point of View of Communications Engineering*, Dover Publications, New York, 1959.

[6] Box, G.E.P. and Muller, M.E., A note on the generation of random normal deviates, *Ann. Math. Statist.*, **29** (1958), 610–11.

[7] Brown, W.G., History of Rand's random digits – summary, Monte Carlo method, *U. S. Nat. Bur. Stand. Appl. Math. Ser.*, No. 12 (1951), 31–2, Government Printing Office.

[8] Brownlee, J., Some experiments to test the theory of goodness of fit, *J. Roy. Statist. Soc.*, **87** (1924), 76–82.

[9] Buslenko, N.P. and Shreider, Yu. A., The method of statistical experiments (the Monte Carlo method) and its practical realization in digital computers, *Fizmatgiz* (1961).

[10] Butcher, J.C., Random sampling from the normal distribution, *Computer J.*, **3** (1961), 251–3.

[11] Butcher, J.C., A partition test for pseudo-random numbers, *Math. Comp.*, **15** (1961), 198–9.

[12] Certaine, J., On sequences of pseudo-random numbers of maximal length, *J. Assoc. Comp. Mach.* **5** (1958), 353–6.

[13] Clark, C.E., The utility of statistics of random numbers, *J. Oper. Res. Soc. Amer.*, **8** (1960), 185–95.

[14] Cochran, W.G., *Sampling Techniques*, John Wiley and Sons, 2nd edn, 1963.

[15] Coveyou, R.R., Serial correlation in the generation of pseudo-random numbers, *J. Assoc. Comp. Mach.*, 7 (1960), 70–74.

[16] Craig, C.C., An application of Thule's semi-invariants to sampling problems, *Metron*, 7 (1928), 3–74.

[17] Cramer, H., *Mathematical Methods of Statistics*, Princeton Univ. Press, 1946.

[18] Curtiss, J.H. (ed.) Monte Carlo method, *U.S. Nat. Bur. Stand. Appl. Math. Ser.*, No. 12 (1951).

[19] Curtiss, J.H., A theoretical comparison of the efficiencies of two classical methods and a Monte Carlo method for computing one component of the solution of a set of linear algebraic equations, *Symposium on Monte Carlo Methods*, ed. H.A. Meyer, 191–233, John Wiley and Sons, 1956.

[20] Cutkosky, R.E., A Monte Carlo method for solving a class of integral equation, *J. Res. Nat. Bur. Stand.* 47 (1951), 113–15.

[21] Davenport, W.B., Jr. and Root, W.L., *An Introduction to the Theory of Random Signals and Noise*, McGraw-Hill, 1958.

[22] Dodd, E.L., Certain tests for randomness applied to data grouped into small sets, *Econometrica*, **10** (1942), 249–57.

[23] Doob, J.L., *Stochastic Processes*, John Wiley and Sons, 1953.

[24] Durbishire A.D., Some tables for illustrating statistical correlations, *Manchester Memoirs*, **57** (1906), No. 16.

[25] Feller, W., *An Introduction to Probability Theory and its Applications*, **2** , 61, John Wiley and Sons, 1966.

[26] Fieller, E.C. and Hartley, H.O., Sampling with control variables, *Biometrika*, **41** (1959), 494.

[27] Forsythe, G.E. and Leibler, R.A., Matrix inversion by a Monte Carlo method, *Math. Tab. Aids Comput.*, **4** (1950), 127–9.

[28] Forsythe, G.E., Generation and testing of random digits, *U.S. Nat. Bur. Stand. Appl. Math. Ser.*, No. 12 (1951), 34–5.

[29] Franklin, J.N., On the equidistribution of pseudo-random numbers, *Quart. Appl. Math.*, **16** (1958), 183–8.

[30] Franklin, J.N., Deterministic simulation of random processes, *Math. Comp.*, **17** (1963), 28–59.

[31] Franklin, J.N., The covariance matrix of a continuous autoregressive vector time-series, *Ann. Math. Statist.*, **34** (1963), 1259–64.

[32] Franklin, J.N., Numerical simulation of stationary and non-stationary Gaussian random processes, *SIAM Rev.*, 7 (1965), 68–80.

[33] Gannon, L.H. and Schmittroth, L.A., Computer generation and testing of random numbers, IDO–16921, A.E.C. Research and Development Report, *Mathematics and Computers*, TID–4500 (22nd edn), 1963.

[34] Gantmacher, F.R., *The Theory of Matrices*, **1** and **2**, transl. by K.A. Hirsch, Chelsea, New York, 1959.

[35] Good, I.J., The serial test for sampling numbers and other tests for randomness, *Proc. Camb. Phil. Soc.*, **49** (1953), 276–84.

[36] Good, I.J., On the serial test for random sequences, *Ann. Math. Statist.*, **28** (1957), 262.

[37] Green, B.F., Smith, J.E.K. and Klem, L., Empirical tests of an additive random generator, *J. Assoc. Comp. Mach.*, **6** (1959) 527–37.

[38] Greenberger, M., Notes on a new pseudo-random number generator, *J. Assoc. Comp. Mach.*, **8** (1961), 163–7.

[39] Greenberger, M., An *a priori* determination of serial correlation in computer generated random numbers, *Math. Comp.*, **15** (1961) 383–9.

[40] Greenberger, M., Method in randomness, *Comm. Assoc. Comp. Mach.*, **8** (1965), 177–9.

[41] Greenwood, M., and White, J.D.C., A biometric study of phagocytosis with special reference to the "opsonic index". Second memoir, on the distribution of the means of samples, *Biometrika*, **7** (1909), 505–30.

[42] Gruenberger, F., Tests of random digits, *Math. Tab. Aids Comp.*, **5** (1950), 244–5.

[43] Gruenberger, F. and Mark, A.M., The d^2 test of random digits, *Math. Tab. Aids Comp.*, **5** (1951), 109–10.

[44] Hall, A., On an experimental determination of π, *Messeng. Math.*, **2** (1873), 113–14.

[45] Halton, H.H., On the efficiency of certain quasi-random sequences of points in evaluating multidimensional integrals, *Numerische Math.*, **2** (1960), 84–90.

[46] Hamaker, H.C., Random sampling numbers, *Statistica Ryswijk*, **2** (1948) 97–106.

[47] Hamaker, H.C., Random frequencies: an expedient for the construction of artificial samples of large size, *Statistica Ryswijk*, **2** (1948), 129–37.

[48] Hamaker, H.C., A simple technique for producing random sampling numbers, *Kon. Ned. Akad. Wet., Proc.*, **52** (1949), 145–50.

[49] Hamaker, H.C., Random sampling frequencies; an implement for rapidly constructing large-size artificial samples, *Kon. Ned. Akad. Wet., Proc.*, **52** (1949), 432–9.

[50] Hammer, P.C., The mid-square method of generating digits, *U.S. Nat. Bur. Stand. Appl. Math. Ser.*, No. **12** (1951), 33.

[51] Hammersley, J.M. and Morton, K.W., Poor man's Monte Carlo, *J. Roy. Statist. Soc.*, (B), **16** (1954), 23–38.

[52] Hammersley, J.M., Monte Carlo method for solving multivariable problems, *Ann. New York Acad. Sci.*, **86** (1960), 844–74.

[53] Hammersley, J.M. and Handscomb, D.C., *Monte Carlo Methods*, Methuen, London, 1964.

[54] Haselgrove, C.B., A method for numerical integration, *Math. Comp.*, **15** (1961), 323–37.

[55] Hicks, J.S. and Whelling, R.F., An efficient method for generating uniformly distributed points on the surface of an n-dimensional sphere, *Comm. Assoc. Comp. Mach.*, **2** (1959), 17–19.

[56] Hull, T.E. and Dobell, A.R., Random number generators, *Soc. Indust. Appl. Math. Rev.*, **4** (1962), 230–54.

[57] International Business Machine Corporation, Random number generation testing, Manual C20–8011, New York, 1959.

[58] Juncosa, M.L., Random number generation on the BRL High Speed Computing Machines, Ballistics Research Laboratories, Report No. 855, Aberdeen Proving Grounds, Maryland, 1953.

[59] Kahn, H. and Harris, T.E., Estimation of particle transmission by random sampling, *U.S. Nat. Bur. Stand. Appl. Math. Ser.*, No. 12 (1951), 27–30.

[60] Kahn, H. and Marshall, A.W., Methods of reducing sample size in Monte Carlo computations, *J. Oper. Res. Soc. Amer.*, **1** (1953), 263–71.

[61] Kahn, H., Multiple quadrature by Monte Carlo methods, in *Mathematical Methods for Digital Computers*, ed. A. Ralston and H.S. Wilf, **1**, 249–57, John Wiley and Sons, 1960.

[62] Kendall, M.G. and Babington Smith, B., Randomness and random sampling numbers, *J. Roy. Statist. Soc.*, **(A) 101** (1938), 147–66.

[63] Kendall, M.G. and Babington Smith, B., Second paper on random sampling numbers, *J. Roy. Statist. Soc.*, **(B) 6** (1939), 51–61.

[64] Kendall, M.G. and Babington Smith, B., *Tracts for Computers*, No. XXIV, Cambridge Univ. Press (1939).

[65] Kendall, M.G. and Stuart, A., *The Advanced Theory of Statistics*, Charles Griffin & Co. Ltd., London, Vol. 1 (1958–69), Vol. 2 (1961–7), Vol. 3 (1966–8).

[66] Lehmer, D.H., Mathematical methods in large-scale computing units, *Ann. Comp. Lab. Harvard Univ.*, **26** (1951), 141–6.

[67] MacLaren, M.D., Marsaglia, G. and Bray, T.A., A fast procedure for generating exponential random variables, *Comm. Assoc. Comp. Mach.*, **7** (1964), 298–300.

[68] MacLaren, M.D. and Marsaglia, G., Uniform random number generators, *J. Assoc. Comp. Mach.*, **12** (1965), 83–9.

[69] Marsaglia, G., Expressing a random variable in terms of uniform random variables, *Ann. Math. Statist.*, **32** (1961), 894–8.

[70] Marsaglia, G., Generating exponential random variables, *Ann. Math. Statist.*, **32** (1961), 899–900.

[71] Marsaglia, G., MacLaren, M.D. and Bray, T.A., A fast procedure for generating normal random variables, *Comm. Assoc. Comp. Mach.*, **7** (1964), 4–10.

[72] Marshall, A.W., The small sample distribution of nW^2, *Ann. Math. Statist.*, **29** (1958), 307.

[73] Mathur R.K., A note on the Wilson–Hilferty transformation of χ^2, *Bull. Calcutta Stat. Assoc.*, **10** (1961), 103–5.

[74] Matuska, W.A. and Innis, G.S., Generation of a stationary Gaussian random process with a specified power spectral density function, Defense Research Laboratory Acoustical Report No. 258, Austin, Texas, 1966.

[75] Meyer, H.A. (ed.), *Symposium on Monte Carlo Methods*, John Wiley and Sons, 1956.

[76] Mood, A.M. and Graybill, F.A., *Introduction to the Theory of Statistics* (2nd edn), McGraw-Hill, 1963.

[77] Moore, P.G., A sequential test for randomness, *Biometrika*, **40** (1953), 111–15.

[78] Moshman, J., The generation of pseudo-random numbers on a decimal calculator, *J. Assoc. Comp. Mach.*, **1** (1954), 88–91.

[79] Moshman, J., Random number generation, in *Mathematical Methods for Digital Computers*, A. Ralston and H.S. Wilf, **2**, 249–63, John Wiley and Sons, 1967.

[80] Muller, M.E., Some continuous Monte Carlo methods for the Dirichlet problem, *Ann. Math. Statist.*, **27** (1956), 569–89.

[81] Muller, M.E., A note on a method for generating points uniformly on N-dimensional spheres, *Comm. Assoc. Comp. Mach.*, **2** (1959), No. 4, 19–20.

[82] Muller, M.E., A comparison of methods for generating normal deviates, *J. Assoc. Comp. Mach.*, **6** (1959), 376–83.

[83] Odell, P.L. and Feiveson, A.G., A numerical procedure to generate a sample covariance matrix, *J. Amer. Statist. Assoc.*, **61** (1966), 199–203.

[84] Page, E.S., The Monte Carlo solution of some integral equations, *Proc. Camb. Phil. Soc.*, **50** (1954), 414–25.

[85] Page, E.S., Pseudo-random elements for computers, *Appl. Statist.*, 8 (1959), 124–31.

[86] Parzen, E., *Stochastic Processes*, Holden-Day, Inc., San Francisco, 1962.

[87] Peach, P., Bias in pseudo-random numbers, *J. Amer. Statist. Assoc.*, **56** (1961), 610–18.

[88] Pearson, E.S. and Hartley, H.O. (ed.), *Biometrika Tables for Statisticians*, **1**, Cambridge Univ. Press, 1954.

[89] Peatman, J.G., and Shafer, R., A table of random numbers from selective service numbers, *J. of Psychology*, **14** (1942), 295–305.

[90] Quenouille, M.H., Tables of random observations from standard distributions, *Biometrika*, **46** (1959), 178–202.

[91] Rand Corporation, *A Million Random Digits with 100,000 Normal Deviates*, Free Press, Glencoe, Illinois, 1955.

[92] Sebestyn, G.S., *Decision-making Processes in Pattern Recognition*, Macmillan Co., New York, 1962.

[93] Sheppard, W.F., The fit of a formula for discrepant observations, *Phil. Trans. Roy. Soc. London*, Ser. A, **228** (1928), 115–50.

[94] Shreider, Yu. A. (ed.), *Method of Statistical Testing (Monte Carlo Method)*, Elsevier Publishing Co., Amsterdam, London and New York, 1964.

[95] Shubick, M., Bibliography on Simulation, Gaming, Artificial Intelligence, and Applied Topics, *J. Amer. Statist. Assoc.*, **55** (1960), 736–51.

[96] Spanier, J. and Gelbard, E.M., *Monte Carlo Principles and Neutron Transport Problems*, Addison-Wesley, Reading, Mass. and London, 1969.

[97] "Student", The probable error of a mean, *Biometrika*, 6 (1908), 1–25.

[98] "Student", Probable error of a correlation coefficient, *Biometrika*, **6** (1908), 302–10.

[99] Taussky, O. and Todd, J., Generation and testing of pseudo-random numbers, in *Symposium on Monte Carlo Methods*, ed. H.A. Meyer, 15–28, John Wiley and Sons, 1956.

[100] Teichroew, D., A history of distribution sampling prior to the era of the computer and its relevance to simulation, *J. Amer. Statist. Assoc.* (Mar, 1965), 26–49.

[101] Thomson, W.E., ERNIE – a mathematical and statistical analysis, *J. Roy. Statist. Soc.*, **A, 122** (1959), 301–33.

[102] Tippett, L.H.C., Random sampling numbers, *Tracts for Computers*, No. XV, Cambridge Univ. Press, 1925.

[103] Tocher, K.D., *The Art of Simulation*, English Universities Press, London, 1963.

[104] Ulam, S., On the Monte Carlo method, *Proc. 2nd Symp. Large-scale Digital Calculating Machinery* (1951), 207–12.

[105] Von Neumann, J., Various techniques used in connection with random digits, *U.S. Nat. Bur. Stand. Appl. Math. Ser.*, No. 12 (1951), 36–8.

[106] Walsh, J.E., Concerning compound randomization in the binary system, *Ann. Math. Statist.*, **20** (1949), 580–9.

[107] Wasow, W., A note on the inversion of matrices by random walks, *Math. Tab. Aids Comp.* **6** (1952), 79–81.

[108] Weldon, W.F.R., Inheritance in animals and plants, in *Lectures on the Method of Science*, ed. T.B. Strong, Clarendon Press, Oxford, 1908.

[109] West, J.H., An analysis of 162,332 lottery numbers, *J. Roy. Statist. Soc.*, **118** (1955), 417–26.

[110] Wilks, S.S., *Mathematical Statistics*, John Wiley and Sons, 1962.

[111] Wilson, E.B. and Hilferty, M.M., The distribution of chi-square, *Proc. Nat. Acad. Sci.*, U.S.A., **17** (1931), 684–8.

[112] Wold, H., Random normal deviates, *Tracts for Computers*, No. XXV, Cambridge Univ. Press, 1948.

[113] Young, L.C., On randomness in ordered sequences, *Ann. Math. Statist.*, **12** (1941), 293–300.

[114] Yule, G.U., On the application of the chi-square method to association and contingency tables, with experimental illustrations, *J. Roy. Statist. Soc.*, **85** (1922), 95–104.